Stephanie Schievenbusch

AAV-mediated gene therapy for renal tubulointerstitial fibrosis

Stephanie Schievenbusch

AAV-mediated gene therapy for renal tubulointerstitial fibrosis

HGF as a multifunctional anti-fibrotic agent with high impact on gene therapy for renal tubulointerstitial fibrosis

Südwestdeutscher Verlag für Hochschulschriften

Impressum/Imprint (nur für Deutschland/ only for Germany)

Bibliografische Information der Deutschen Nationalbibliothek: Die Deutsche Nationalbibliothek verzeichnet diese Publikation in der Deutschen Nationalbibliografie; detaillierte bibliografische Daten sind im Internet über http://dnb.d-nb.de abrufbar.

Alle in diesem Buch genannten Marken und Produktnamen unterliegen warenzeichen-, marken- oder patentrechtlichem Schutz bzw. sind Warenzeichen oder eingetragene Warenzeichen der jeweiligen Inhaber. Die Wiedergabe von Marken, Produktnamen, Gebrauchsnamen, Handelsnamen, Warenbezeichnungen u.s.w. in diesem Werk berechtigt auch ohne besondere Kennzeichnung nicht zu der Annahme, dass solche Namen im Sinne der Warenzeichen- und Markenschutzgesetzgebung als frei zu betrachten wären und daher von jedermann benutzt werden dürften.

Verlag: Südwestdeutscher Verlag für Hochschulschriften GmbH & Co. KG
Dudweiler Landstr. 99, 66123 Saarbrücken, Deutschland
Telefon +49 681 37 20 271-1, Telefax +49 681 37 20 271-0
Email: info@svh-verlag.de
Zugl.: Köln, Universität zu Köln, Diss., 2009

Herstellung in Deutschland:
Schaltungsdienst Lange o.H.G., Berlin
Books on Demand GmbH, Norderstedt
Reha GmbH, Saarbrücken
Amazon Distribution GmbH, Leipzig
ISBN: 978-3-8381-1991-5

Imprint (only for USA, GB)

Bibliographic information published by the Deutsche Nationalbibliothek: The Deutsche Nationalbibliothek lists this publication in the Deutsche Nationalbibliografie; detailed bibliographic data are available in the Internet at http://dnb.d-nb.de.

Any brand names and product names mentioned in this book are subject to trademark, brand or patent protection and are trademarks or registered trademarks of their respective holders. The use of brand names, product names, common names, trade names, product descriptions etc. even without a particular marking in this works is in no way to be construed to mean that such names may be regarded as unrestricted in respect of trademark and brand protection legislation and could thus be used by anyone.

Publisher: Südwestdeutscher Verlag für Hochschulschriften GmbH & Co. KG
Dudweiler Landstr. 99, 66123 Saarbrücken, Germany
Phone +49 681 37 20 271-1, Fax +49 681 37 20 271-0
Email: info@svh-verlag.de

Printed in the U.S.A.
Printed in the U.K. by (see last page)
ISBN: 978-3-8381-1991-5

Copyright © 2010 by the author and Südwestdeutscher Verlag für Hochschulschriften GmbH & Co. KG and licensors
All rights reserved. Saarbrücken 2010

Table of Contents

TABLE OF CONTENTS	**I**
1. INTRODUCTION	**1**
1.1 RENAL FAILURE	1
1.1.2 Interstitial fibrosis	1
1.1.3 Mouse models of interstitial fibrosis in the kidney	2
1.1.3.1 COL4A3 knockout mouse	3
1.2 TGF β: THE KEY MEDIATOR OF MATRIX ACCUMULATION	4
1.2.1 Signal transduction by TGFβ	5
1.2.2 Smad proteins and their assembly	6
1.3 HEPATOCYTE GROWTH FACTOR (HGF)	7
1.3.1 HGF signaling	8
1.4 HGF AND TGFβ IN RENAL FIBROSIS	10
1.5 GENE DELIVERY SYSTEMS TO TARGET CHRONIC KIDNEY DISEASES	10
1.5.1 Recombinant AAV as a gene delivery vector	11
1.5.2. Characteristic properties of AAV	12
1.5.3 Production of recombinant AAV vectors	13
1.5.4 Different AAV serotypes	14
1.5.5 Pseudo-packaging	15
1.5.6 Gene delivery to the kidney	16
1.5.7 Clinical trials/therapeutical approaches of AAV	16
1.6 AIM OF THE STUDY	18
ABBREVIATIONS	**19**
2. MATERIALS AND METHODS	**22**
2.1 MATERIALS	22
2.1.1 Chemicals, plastic ware and other materials	22

TABLE OF CONTENTS

- 2.1.2 Software 22
- 2.1.3 Enzymes and antibodies 22
  - 2.1.3.1 Enzymes 22
  - 2.1.3.2 Antibodies 23
  - 2.1.3.3 Consumables 23
  - 2.1.3.4 Devices 24
  - 2.1.3.5 Cell culture 24
  - 2.1.3.6 Reagents 24
  - 2.1.3.7 Cytokines 24
- 2.1.4 Kits and assays 25
- 2.1.5 Oligonucleotides 25
- 2.1.6 Plasmids 28
- 2.1.7 Buffers and solutions 29
- 2.1.8 Cell lines 30

2.2 METHODS 31
- 2.2.1 DNA preparation 31
  - 2.2.1.1 DNA preparation of *E. coli* 31
  - 2.2.1.2 DNA extraction by phenol chloroform extraction 31
  - 2.2.1.3 DNA extraction from mouse tails 32
- 2.2.2 RNA preparation 32
  - 2.2.2.1 RNA isolation 32
- 2.2.3 DNA modification 32
  - 2.2.3.1 Restriction analysis 32
  - 2.2.3.2 Dephosphorylation by alkaline phosphatase 33
  - 2.2.3.4 Generation of blunt ends 33
  - 2.2.3.5 Dimerisation and phosphorylation of oligonucleotides 33
  - 2.2.3.6 Ligation of DNA fragments 34
- 2.2.4 Transformation 34

TABLE OF CONTENTS

- 2.2.4.1 Preparation of competent bacteria 34
- 2.2.5 Polymerase chain reaction (PCR) 35
  - 2.2.5.1 Qualitative PCR 35
  - 2.2.5.2 Genotyping of mice 36
  - 2.2.5.3 Real-time PCR 37
  - 2.2.5.4 Determination of the titer of AAV preparations or the AAV infection rate by quantitative PCR 38
  - 2.2.5.5 Sequencing of DNA 39
  - 2.2.5.6 Reverse transcriptase reaction 39
- 2.2.6 Biochemical methods 40
  - 2.2.6.1 Preparation of protein extracts and determination of protein concentration 40
  - 2.2.6.2 Western blot analysis 40
- 2.2.7 Cell culture 41
  - 2.2.7.1 Cell culture medium 41
  - 2.2.7.2 Passage of cells 41
  - 2.2.7.3 Cryopreservation of cells 41
  - 2.2.7.4 Stimulation of cells 42
  - 2.2.7.5 Transfection of plasmid DNA 42
  - 2.2.7.6 HGF ELISA 43
- 2.2.8 Functional Analysis 43
  - 2.2.8.1 Luciferase Assay 43
- 2.2.9 AAV production 43
  - 2.2.9.1 Preparation of AAV 43
  - 2.2.9.2 AAV extraction and purification 44
- 2.2.10 The COL4A3 knockout mouse model 45
  - 2.2.10.1 Transduction of mice with AAV 45
  - 2.2.10.2 Preparation of organs from adult mice 45
  - 2.2.10.3 Fixation, paraffin embedding and microtoming of mouse organs 45

2.2.10.4 Morphological and immunohistochemical studies 46

2.2.10.5 Classification of the fibrosis grade in transversal kidney sections 46

2.2.10.6 Immunohistochemical staining for α-SMA 46

2.2.10.7 Immunohistochemical staining for GFP 47

3. RESULTS 47

3.1 HGF ACTING AS AN ANTI-FIBROTIC AGENT 47

3.1.1 HGF stimulates the Erk1/2 pathway and the Akt pathway in renal fibroblasts......... 47

3.1.2 Expression profiles induced by hHGF stimulation in renal fibroblasts..................... 50

3.1.2.1 HGF inhibits expression of signal transducers linked to fibrotic processes 53

3.1.2.2 Effects of HGF on the expression level of Smad genes.. 54

3.1.2.3 Effects of HGF on the expression level of CCN family members........................ 55

3.1.2.4 Effects of HGF on the expression level of fibrotic markers and collagens 55

3.1.2.5 Smad independent anti-fibrotic effects of HGF.. 56

3.2 THE ANTI-FIBROTIC FUNCTION OF HGF IN A GENE THERAPEUTICAL APPROACH, *IN VIVO* 60

3.2.1 Renal transduction efficiency of different AAV serotypes 61

3.2.1.1 Low *in vitro* transduction of renal epithelial cells by rAAV2 61

3.2.1.2 *In vivo* transduction of liver and kidney by rAAV2, rAAV8 and rAAV9............ 62

3.2.2 Transgene expression limited to renal tubuloepithelial cells..................................... 66

3.2.2.1 Transgene expression driven by the Ksp-promoter in renal tubuloepithelial cells *in vitro* 67

3.2.2.2 Reporter gene expression driven by the kidney-specific Ksp-promoter *in vivo* ... 69

3.2.3 AAV-induced recombinant hHGF expression as a gene therapeutical approach for the treatment of tubulo-interstitial fibrosis *in vivo* 71

3.2.3.1 Construction of the hHGF-expression cassette for the generation of AAV8 and 9 vectors... 71

3.2.3.2 *In vivo* administration of hHGF by rAAV8 and rAAV9 72

3.2.4 Anti-fibrotic function of AAV-mediated hHGF expression 74

3.2.4.1 Repression of fibrotic markers by recombinant hHGF expression 74

3.2.3.6 Deceleration of fibrotic remodelling of the kidney architecture 77

4. DISCUSSION 79

4.1 HGF MEDIATES ITS ANTI-FIBROTIC EFFECTS BY MAP/ERK- AND BY AKT-SIGNALING

4.2 PROFILING OF ANTI-FIBROTIC SIGNALS IN INTERSTITIAL FIBROBLASTS 80

4.3 A GENE THERAPEUTICAL APPROACH TARGETING HGF TO RENAL FAILURE, *IN VIVO* 86

4.3.1 HGF applied by the adeno-associated virus *in vivo* 87

4.3.2 AAV9 is superior to AAV8 and AAV2 in kidney and liver transduction 87

4.3.3 Choice of a specific promoter 88

4.3.4 HGF mediates anti-fibrotic effects identified *in vitro* also *in vivo* 89

4.3.5 Paracrinal and endocrinal delivery of hHGF 92

4.4 PERSPECTIVES OF ANTI-FIBROTIC FUNCTIONS BY AAV MEDIATED HGF TRANSFER IN GENE THERAPEUTICAL APPROACHES 95

5. REFERENCES 96

6. SUPPLEMENTS 107

1. Introduction

1.1 Renal failure

Terminal renal failure is a life-threatening disease pattern only controllable by chronic dialysis or renal transplantation. The incidence is the result of acute or chronic disease processes and based on hereditary dysplasia or dysfunction of the kidney, respectively. In the last decade prevalence and incidence of chronic renal failure in Western Europe and USA continuously accelerated with more than 150.000 new cases per year in USA, predominantly driven by the rise of obesity, hypertension and diabetes in ageing Western populations [1, 2]. The overall rate of new cases due to diabetes was 50 % higher than a decade ago. Also the cases due to hypertension increased about 48 % compared to 1996. The incidence rate for dialysis in the USA accounted for 352 people per million populations (USRDS, 2008). Patients suffering chronic renal failure thereby require the highest expense per capita health care costs at all.

Among different renal lesions, linked to proteinuria due to dysfunction in glomerular filtration, diabetic nephropathy is the major reason of renal failure. Hyperfiltration and proteinuria lead to thickening of the glomerular basement membrane and inflammatory alterations, followed by tubulointerstitial fibrosis, which in turn is associated with extracellular matrix accumulation, inflammatory cell infiltration and myofibroblastic precursor transdifferentiation and proliferation in the tubulointerstitium [3-5].

1.1.2 Interstitial fibrosis

The progressive loss of renal function is both associated with development of glomerulosclerosis and interstitial fibrosis [6]. In general, fibrosis is the end result of chronic inflammation induced by a variety of stimuli including persistent infections, autoimmune reactions, toxins, allergic responses, radiation, and tissue injury [7]. The formation of renal interstitial fibrosis is basically a multifunctional event that can be separated in four phases [8-10]: (1) cellular activation and injury, (2) transmission of fibrogenic signals, (3) formation of fibrosis, and finally (4) destruction of the kidney. In the first step renal tubules become activated and an interstitial infiltration of monocytes/macrophages and T-cells occurs. In the second step these cells release soluble factors like several growth factors and cytokines with pro-fibrotic effects (e.g. transforming growth factor beta (TGFβ), connective tissue growth factor (CTGF), angiotensin II and endothelin-1). The third step includes the fibrogenic phase. Histological this incident is characterized by an accumulation of extracellular matrix in the interstitium. In the fourth and last phase the tubules and peritubular capillaries are destructed and the glomerular filtration is reduced caused by progressively declining nephrons.

The central cell types of fibrosis are the myofibroblastic cells that are regarded as contractile intermediates of fibroblasts and smooth muscle cells. Their origin, however, can vary between various cell types in different organs [7, 11]. Thus, during renal tubulointerstitial fibrosis myofibroblasts originate mainly from interstitial fibroblasts and even from tubular epithelial cells, although to a much lesser extent. The transformation of epithelial cells into myofibroblasts is characterized by the gradual loss of the morphological and functional integrity of the tubular epithelium, cytoskeletal alterations and the deposition of extracellular matrix (ECM) and is summarized as epithelial-to-mesenchymal transition (EMT) [12].

On molecular level the process of renal fibrosis is based on the accumulation of extracellular matrix proteins. These accumulation can be attributed to [13]:

1. increased production of extracellular matrix proteins (e.g. fibronectin, proteoglycanes, collagen I, III, IV, laminin, vitronectin and tenascin)
2. reduced degradation of extracellular matrix by dint of diminished synthesis of proteases (e.g. metalloproteases like MMP-1, collagenase I, gelatinase A and B)
3. overexpression of matrix-binding receptors (e.g. integrins)

1.1.3 Mouse models of interstitial fibrosis in the kidney

While there is a diversity of primary pathomechanisms that lead to renal failure, all chronic kidney diseases finally result in the development of interstitial fibrosis. To date there is a variety of genetic and inducible animal models mimicking renal diseases, however only few of them progress consistently to interstitial fibrosis [14]. For example the model of unilateral ureter obstruction (UUO) represents an inducible model of tubulointerstitial fibrosis. In this model the ureter of one kidney becomes ligated, while the contra-lateral kidney serves as control. This ligature results in severe fibrosis, however the progression of disease is very unphysiological as it develops within 2 weeks. In contrast, a physiological mouse model which is similar to the human disease of Alport syndrome is the COL4A3 knockout mouse model.

1.1.3.1 COL4A3 knockout mouse

One cause of chronic kidney disease is the Alport syndrome were type IV collagen, one compound of the glomerular basement membrane (GBM), is absent or abnormal. The GBM is a key component of the blood filtration apparatus in the kidney, formed by assembly of type IV collagen with nidogen, laminins, and proteoglycans [15]. In the Alport syndrome the glomeruli become inflamed and scarred, and slowly lose their ability to remove waste and excess water from the blood. The COL4A3 knockout mouse model, developed by Cosgrove et al., is one of several animal models of the Alport syndrome and represents the autosomal form. These mice lack expression of the collagen chains α3(IV), α4(IV), and α5(IV) [16]. The disease-progression is very similar to that reported in studies of humans and include hearing defects, microhematuria, proteinuria, and irregular thickening and splitting of the glomerular basement membrane (GBM) [16]. Microhematuria starts with the age of two weeks and progresses until the death of the animals. With the age of five weeks proteinuria is detectable and the protein content raises rapidly until 6 - 6.5 weeks (10 – 15 mg/ml) to remain constant until the end. Mice suffering from the Alport syndrome finally develop an interstitial fibrosis and based on the genetic background (129/SvJ) these mice develop endstage renal failure within 14 weeks [16]. The morphology of the kidney is characterized by thickening and splitting of the GBM. The thickening starts with the age of four weeks in the external capillary loops and spreads out in the whole kidney with the age of eight weeks. With fourteen weeks half of the glomeruli are fibrotic, the capillaries are collapsed and the kidney is 30 – 50 % smaller compared to the wildtype [16].

To date, Alport syndrome is treated by ACE-inhibitors, dialysis and renal transplantation. In general, the ultimate therapy for end-stage renal failure is dialysis or transplantation. However, these treatments implicate medical and social disadvantages. Furthermore, they are not always successful. Therefore, novel therapeutical strategies are still hiughly required.

1.2 TGF β: The key mediator of matrix accumulation

Transforming growth factor beta (TGFβ) plays a decisive role in wound-healing and tissue repair and is currently viewed as the main mediator of fibrotic processes and responsible for enhanced synthesis of extracellular matrix proteins by mesenchymal cell types. It is a multifunctional cytokine exhibiting diverse biological effects in cellular processes including proliferation, migration, differentiation, and apoptosis [17]. TGFβ takes the key role by the expansion of extracellular renal matrix by its influence on the three molecular mechanisms already mentioned: an increased production of extracellular matrix that is accompanied by a decreased degradation of the renal matrix, based on down-regulated gene expression of matrix-degrading enzymes [18].

The TGFβ superfamily consists of more than 60 distinct ligands that include TGFβ, activins and bone morphogenic proteins (BMPs) [17]. Three isoforms of TGFβ have been identified (TGFβ1, -2 and -3) whereas TGFβ1 is the most extensive investigated [10] and considered to be the major or predominant isoform involved in renal fibrosis [19]. The roles of TGFβ2 and TGFβ3 are considerably more unclear, however, all three TGFβ isoforms are known to be involved in matrix synthesis and degradation [19]. Most TGFβ that is isolated from plasma, urine or renal tissue exists in an inactive form, the latent precursor (LTGFβ) [20]. This precursor consists of the N-terminal pro-domain with the latency associated peptide (LAP) and the C-terminal potentially bioactive region, the mature TGFβ [20, 21]. The activation into the mature form for receptor binding is mediated either by proteolytic cleavage, interaction of the LAP region with other proteins, reactive oxygen species or low pH [20, 22]. The receptors for TGFβ are the receptors type I, II, and III. The type III receptor is a membrane anchored proteoglycan that is also called betaglycan. This receptor has no signaling structure, does not bind the ligand and is assumed to play an indirect role in TGFβ signaling [17]. On the contrary, type I and type II receptors are transmembrane serine-threonin kinases that mediate the TGFβ signaling [20]. In the absence of TGFβ, both are present as homodimers. Upon ligand binding to the type II receptor, this receptor phosphorylates and activates the type I receptor, resulting in a heteromeric ligand-receptor complex [20].

1.2.1 Signal transduction by TGFβ

Various pathways are involved in TGFβ signaling, including signal transduction by the TAK1/p38/JNK and MAPK pathways, the PI3 Kinase/Akt-mTor, or the Rho pathways, but most importantly the Smad pathway [23]. The activation of MAPK, TAK1/p38/JNK and PI3K play a major role in the regulation of EMT [23] while the onset of TAK1/p38/JNK is also linked to apoptosis [24, 25]. Furthermore, proliferation of fibroblasts and morphological transformation can be attributed to the activation of PI3K via TGFβ [26], whereas the induction of stress fiber formation and mesenchymal characteristics in epithelial cells are reported to be mediated via RhoA [23]. These pathways, although they can act in conjunction with the Smad-pathway, are Smad-independent.

However, the best-known pathway for TGFβ signaling, also in connection with fibrosis, is the Smad-dependent pathway. TGFβ signaling via this cascade involves membrane receptors and Smad-transcription factors. The existing eight different Smad proteins can be divided into three functional groups: (1) receptor-regulated Smads (R-Smads), (2) co-mediator Smads (co-Smads), and (3) the inhibitory Smads (I-Smads) [27].

Smad 1, 2, 3, 5 and 8 belong to the R-Smads. While Smad1, 5, and 8 are mediators of activin and BMP signals, Smad2 and 3 are responsible for TGFβ signaling. After TGFβ binding to its TGFβ type II receptor, this serine/threonine kinase subsequently recruits and phosphorylates the TGFβ type I receptor (Fig. 1.1). The arising ligand-heterodimeric receptor complex in turn phosphorylates the receptor-associated cytoplasmic mediators Smad2 or Smad3 on a C-terminal SSXS motif [28, 29]. Once activated, they interact with the common Smad4 and translocate as a heteromeric complex into the nucleus where they activate target genes by either direct DNA binding or in association with other transcription factors [18]. In contrast, members of the third functional group, the inhibitory Smads (Smad6 and Smad7), negatively regulate TGFβ signaling by inhibition of phosphorylation and/or nuclear translocation of R-Smads [30, 31].

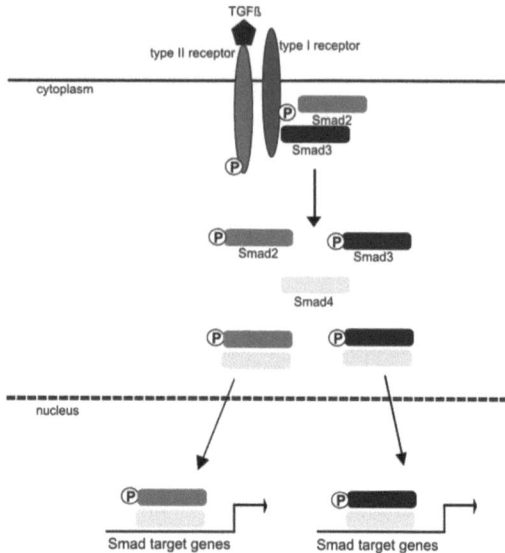

Fig. 1.1: TGFβ signaling. Upon TGFβ binding to its TGFβ type II receptor this serine/threonine kinase becomes activated and subsequently recruits and phosphorylates the TGFβ type I receptor. The arising ligand-heterodimeric receptor complex in turn phosphorylates the receptor-associated cytoplasmic mediators Smad2 or Smad3. Once activated, they form complexes with Smad4 and translocate into the nucleus where they activate target genes by either direct DNA binding or in association with other transcription factors.

1.2.2 Smad proteins and their assembly

The Smad transcription factors activated by TGFβ are Smad2 and Smad3 (R-Smads), Smad4 (co-Smad) and the inhibitory Smad7 [32]. The signaling of Smad2, Smad3 and Smad4 is carried out as already mentioned. The inhibitory Smad7 has been reported to mediate its negative effects via two mechanisms. On the one hand, it inhibits TGFβ signaling by stable binding to the activated type I receptor resulting in a suppressed phosphorylation of Smad2 and Smad3 [33, 34]. On the other hand, Smad7 is able to recruit ubiquitin ligases to the TGFβ receptor complex, leading to the degradation of the type I receptor [35].

Smads are made up of two highly conserved domains, the Mad Homology domains MH1 and MH2 that are connected via a linker region [36], [37]. While the N-terminal MH1 domain interacts with the DNA [38], the C-terminal MH2 domain is equipped with transcriptional activation properties [37] and is responsible for the binding of receptors, partner Smads and co-activators for transcription [29, 39].

1.3 Hepatocyte growth factor (HGF)

Numerous studies reveal that in contrast to the pro-fibrotic TGFβ, hepatocyte growth factor (HGF) can act as an anti-fibrotic factor by directly antagonizing the pro-fibrotic actions of TGFβ [40], [41].

The hepatocyte growth factor (HGF) was discovered in the late 1980s as a unique protein that promotes hepatocyte proliferation and liver regeneration [40-42]. It is a heterodimeric protein that consists of a heavy (54-64 kDA) α-chain and a light (31.5-34.5 kDa) β-chain linked together by a disulfide bond [43]. Both chains are produced from the native inactive single chain precursor HGF (pro-HGF) that becomes activated after extracellular proteolysis by several activators [44-46].

HGF is a multifunctional polypeptide that exhibits mitogenic, motogenic, morphogenic and anti-apoptotic activities [47]. Furthermore, HGF is known to play a decisive role in organ protection and regeneration [47, 48]. With regard to renal diseases an abrupt rise of HGF mRNA and/or protein levels in the kidney were detected in various types of acute renal injury [49-52] indicating that the kidney is one source of HGF. But also in distinct organs like liver, spleen and lung HGF levels increased in case of a renal injury [53-55].

Recent studies have shown that in addition to the role of HGF in organ regeneration, HGF is an anti-inflammatory and anti-fibrotic factor. The anti-fibrotic action of HGF was first observed in experimental liver fibrosis of chronically intoxicated rats [56]. Subsequently, the anti-fibrotic function of locally or systemically applied HGF was demonstrated in a variety of experimental systems including models of murine and porcine renal failure. HGF application resulted in deceleration of histological fibrotic changes and less extracellular matrix deposition [57-59].

The biological activities of HGF are mediated via the specific receptor c-met, a receptor that belongs to the receptor tyrosine kinase superfamily and is the product of the *c-met* protooncogene [60, 61]. This receptor is composed of an N-terminal α-chain located outside the membrane and a C-terminal β-chain containing an intracellular tyrosine kinase domain [41, 62].

1.3.1 HGF signaling

The pleiotropic effects of HGF are mediated by the c-met receptor through different signaling pathways (Fig. 1.2). Upon HGF binding to the receptor kinase c-met, the receptor becomes autophosphorylated at specific tyrosine residues (Tyr-1349 and Tyr 1356) that function in the phosphorylated form as docking sites for multiple signal-transducers and adaptor molecules such as Gab1, Grb2, SOS, PI3K, and Stat [63-67]. These molecules in turn induce the activation of the signal transduction cascades of PI3K/Akt, Ras-Raf-MAPK, and Stat3.

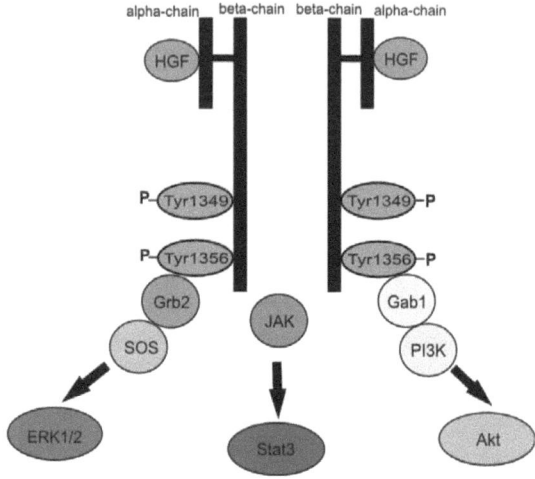

Fig. 1.2: Schematic signal-transduction of HGF via its c-met receptor. C-met is composed of an alpha-chain and a transmembrane beta-chain which are linked by a disulfide bond. Two tyrosine residues in the beta-chain, Tyr1349 and Tyr1356, are necessary for the signal transduction. They constitute docking sites which recruit various signaling and adaptor proteins including Gab1, Grb2, SOS, PI3K and Stat3. These molecules then elicit the activation of the signal transduction pathways MAP/Erk, Stat3 and Akt.

However, the molecular and cellular mechanisms underlying these activities are not well understood. Recently, Yang et al. reported that HGF mediates its anti-fibrotic effects by antagonizing the pro-fibrotic function of TGFβ [57]. Further studies of this group could show that stimulation with HGF inhibited TGFβ signal transduction by

ERK1/2 initiated phosphorylation and blockade of the Smad2/3 transducers in fibroblasts, mesangial cells and epithelial cells [29, 68, 69].

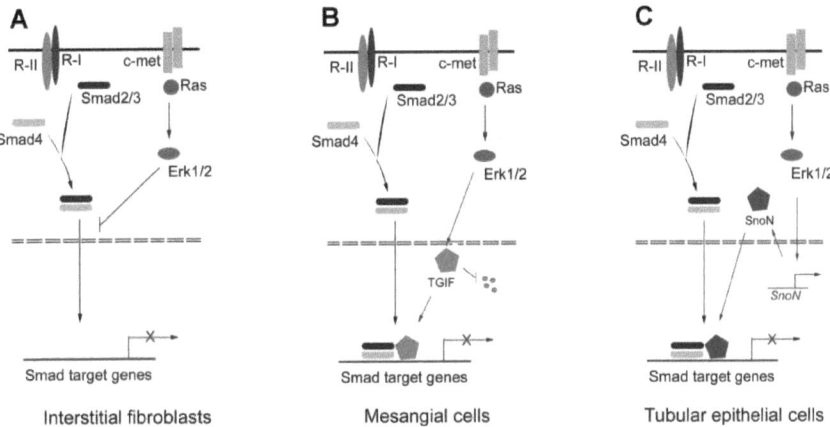

Fig. 1.3: HGF antagonizes TGFβ signaling in various types of kidney cells through the activation of Erk1/2. TGFβ binding to its TGFβ type II receptor (R-II) activates this serine/threonine kinase and subsequently recruits and phosphorylates the TGFβ type I receptor (R-I). The arising ligand-heterodimeric receptor complex in turn phosphorylates the receptor-associated cytoplasmic mediators Smad2 or Smad3. Once activated, they form complexes with Smad4 and translocate into the nucleus where they activate target genes by either direct DNA binding or in association with other transcription factors.
(A) in interstitial fibroblasts, HGF signaling can inhibit TGFβ signal transduction by ERK1/2 initiated phosphorylation that results in the blockade of phosphorylated Smad2/3 nuclear translocation. (B) in mesangial cells, HGF signaling rapidly induces expression of the Smad transcriptional co-repressor TGIF and stabilizes it from degradation. Accumulated TGIF binds to Smads thereby suppressing transcription of TGFβ target genes. (C) in tubular epithelial cells, HGF signaling induces expression of another Smad transcriptional co-repressor, SnoN. As TGIF, SnoN binds to phosphorylated Smads and intercept Smad2/3-mediated gene expression. (Liu, 2004 [74])

In fibroblasts, activation of Erk1/2 results in the phosphorylation of Smad2/3 in the linker region thereby inhibiting their nuclear translocation and the onset of Smad target gene expression (Fig. 1.3 A). On the contrary, in mesangial cells and epithelial cells Erk1/2 activation initiates the transcription of two Smad co-repressors (TGIF or SnoN) that do not block nuclear translocation of activated Smad2/3 but block transcription of Smad-target genes (Fig. 1.3 C, D).

1.4 HGF and TGFβ in renal fibrosis

The anti-fibrotic role of HGF seems to be in close connection to the pro-fibrotic TGFβ. *In vitro* and *in vivo* studies could already show that HGF antagonizes many pro-fibrotic actions of TGFβ and it is assumed that the balance between TGFβ and HGF plays an essential role in the pathogenesis of chronic renal fibrosis [70, 71]. While TGFβ supports renal fibrosis by cellular processes that include (1) apoptosis of podocytes, endothelial- and tubuloepithelial-cells, (2) activation of mesangial cells and interstitial fibroblasts to produce huge amounts of matrix components and (3) initiation of epithelial to mesenchymal transition (EMT), HGF targets these cellular processes that are decisive for renal fibrogenesis. Both growth factors are initially induced after tissue injury and are expected to be necessary and important for the initial wound-healing response and tissue repair [72, 73]. Whereas a transient injurious stimulus will result in a predominant HGF signaling, yielding tissue repair and regeneration, a chronic injury leads to a progressively increasing TGFβ expression that finally will prevail. In case of the latter HGF initially increases but gradually declines [74].

The beneficial effect of exogenous HGF has been shown in different experimental models of chronic kidney diseases [47, 59, 75-77]. For example the onset of tubulointerstitial fibrosis has been shown to be inhibited by the application of recombinant HGF [75]. Also systemic administration of naked plasmid DNA encoding human HGF has been reported to attenuate renal interstitial fibrosis and reduce TGFβ and its type I receptor expression in a mouse model of UUO (unilateral ureteral obstruction) [78]. To date, most of the pre-clinical studies in mice and porcine models of renal failure report of a systemic or local HGF application by recombinant protein or plasmid DNA [57-59, 78-80]. But there is also the possibility to use other systems for the delivery to the organ or tissue of interest.

1.5 Gene delivery systems to target chronic kidney diseases

The idea of gene therapy is the introduction of genetic material into an organism to improve or even cure a disease. In general, there are two different delivery systems that can be used for gene transfer, viral and non-viral vectors. The non-viral vector strategy uses naked DNA or DNA complexed with cationic lipids, polymers or peptides [81] whereas the viral systems include adeno-, retro-, pox-, herpex simplex-, and adeno-associated viral vectors [82, 83]. However, all of these gene transfer systems offer potential advantages and disadvantages. The argue for non-viral vectors is a broad tropism, limited immunogenicity, easy and cost-saving preparation, applicability for multiple treatment and the lack of limitation of the infused DNA [82-84]. The disadvantages of the non-viral delivery systems include an inefficient intracellular processing and a transient gene expression in proliferating tissue due to the lack of genome integration [82].

Retroviruses, in contrast, stably integrate their genome into the host DNA thereby stably expressing the therapeutic gene [82]. Though, an important negative aspect of these viral vectors is the need of actively dividing cells for efficient transduction [82, 85].

On the contrary, adenoviral vectors allow for a transduction of dividing and non-dividing cells. In addition, they are easy to produce at high titers and they offer a large packaging capacity [84]. The disadvantages in case of adenoviral vectors, however, include a transient expression and also a cellular immune response [82, 84]. In addition, a humoral response to the injected vector is often generated, preventing the use of such vectors for repeated administration [86] or the necessity of an additional co-application of immunosuppressive agents or cytokines [86]. A viral vector that might overcome most of the mentioned disadvantages is the recombinant adeno-associated virus (AAV).

1.5.1 Recombinant AAV as a gene delivery vector

In the recent years the adeno-associated viral vectors (AAV) have attracted interest for gene therapy of various genetic disorders [87-90], representing a multiplicity of features that characterize them as an ideal gene vehicle. The main reason is the safety of this vector. No real alterations in the pathology have been observed by AAV infection. In addition, AAV vectors are replication deficient and fail to initiate an immune response. Another advantage of AAV is the capability to transduce both, proliferating and quiescent cells [91, 92]. Moreover, AAV allows for a long-term expression of the transgene by site-specific integration into the genome or episomal persistence [93]. However, a disadvantage is the limited packaging capacity of approximately 4.5 kb [94] and the need for second-strand synthesis which exhibits a rate limiting step for primary cell transduction and *in vivo* application [83].

1.5.2. Characteristic properties of AAV

AAV belongs to the *Dependovirus* genus of the *Parvoviridae* family and is one of the smallest and structurally most uncomplex viruses. The first human adeno-associated virus (AAV) was discovered in 1965, as a contaminant of adenovirus preparations [95]. The linear single-stranded DNA (ssDNA) genome of approximately 4.7-kilobases (kb) can be divided into three functional regions: (1) the terminal inverted repeats (ITRs), (2) the 5' located *rep* open reading frame (ORF) and (3) the 3' located *cap* ORF. The ITR's comprise 145 nt and are the only *cis* elements that are required for replication, packaging, and insertion into the host genome [91]. The two ORFs that are flanked by the ITRs, *rep* and *cap,* encode non-structural and structural proteins, respectively. *Rep* encodes four proteins that are necessary for replication of the viral genome (Rep78, Rep68, Rep52 and Rep40) whereas *cap* encodes the three capsid proteins (VP-1, VP-2 and VP-3) (Fig. 1.4). The replication of AAV is dependent on a helper virus such as adenovirus or herpes simplex virus [82, 96, 97]. In the absence of a helper, AAV persists in a latent form, either by site-specific integration into the human genome at a specific region on chromosome 19 or as episomal form (double stranded circular or linear) [98].

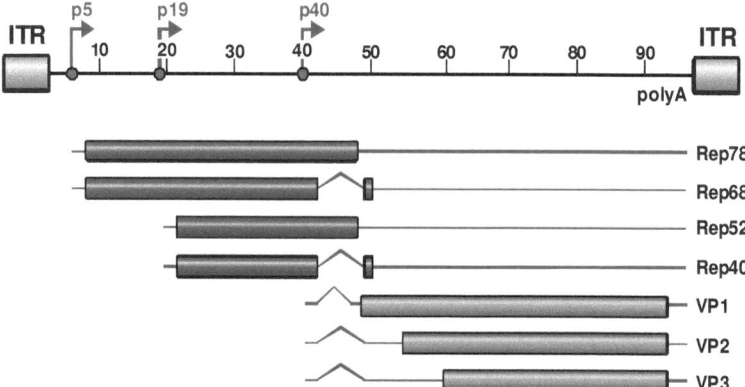

Fig. 1.4: Organisation of the AAV genome and gene products. Shown are the inverted terminal repeats (ITRs), the three viral promoters positioned at units 5, 19 and 40 (p5, p19, p40) and the polyadenylation signal at unit position 96 (poly A). Cylinders represent open reading frames. Untranslated regions are indicated as solid lines and introns as kinks. Expression of the four Rep proteins is regulated by p5 and p19, whereas the p40 promoter controls expression of the three different capsid proteins VP1, VP2 and VP3. (Figure was kindly provided by N. Huttner).

1.5.3 Production of recombinant AAV vectors

Recombinant AAV vectors possess a coding capacity of 4.5 kb [94]. For the generation of recombinant AAV vectors all viral genes (*rep* and *cap*) can be completely deleted. This is due to the fact that only AAV ITRs contain all *cis* acting elements essential for viral replication, packaging and integration. The deleted *rep/cap* sequences of the parental virus can be replaced by a marker or therapeutic gene thereby generating vectors that are replication deficient even in the presence of a helper virus [99]. In general, the vector production is performed in a helper virus-free manner to avoid helper virus contaminations of vector preparations. The *rep* and *cap* gene products as well as the essential adenoviral genes VA, E2A and E4 are supplied by a helper plasmid in *trans* [97, 100]. Therefore, the production requires co-transfection of three plasmids into HEK239 cells that already contain another essential protein, namely the E1A/E1b gene: (1) an AAV-plasmid with the transgene expression cassette flanked by the ITRs, (2) a helper plasmid in *trans* that carries *rep* and *cap* and (3) a plasmid that provides the adenoviral helper genes [101] (Fig. 1.5).

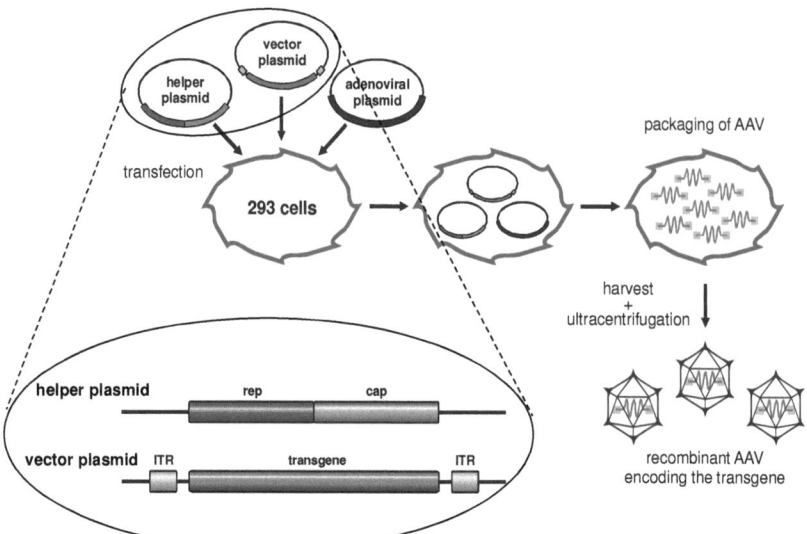

Figure 1.5: Production of recombinant AAV in HEK293 cells. The vector plasmid encodes for the vector genome that is flanked by the inverted terminal repeats (ITR's). Both viral ORFs can be replaced by the transgene. The helper plasmid is responsible for providing the *rep* and *cap* genes. Helper virus functions are encoded on a third plasmid (adenoviral helper plasmid). AAV virions can be harvested 48 h after triple transfection, and are purified by density gradient and/or chromatography. (Figure was kindly provided by H. Büning)

The AAV particles can be harvested 48 h p.i. out of the lysates of the transfected cells and purified to high titers (up to 10^{14} particles/ml) using one of several described protocols [87, 89, 102, 103].

1.5.4 Different AAV serotypes

In addition to the most commonly used rAAV2 numerous new serotypes and variants have been isolated [95, 104-107]. AAV1 to AAV4 and AAV6 were isolated as contaminants in different laboratory adenovirus type stocks [95], [105], [108] whereas AAV5 was isolated from a human penile condylomatous lesion [109]. Serotypes AAV7 and AAV8 were cloned from rhesus monkey tissues [104], while AAV9 was found in human tissue [107]. Further isolates are AAV10 and AAV11 that were isolated from cynomolgus monkeys [110].

It could already be shown *in vitro* and *in vivo* that the various natural AAV serotypes display different tropisms for tissues or cells (Table 1.1).

Table 1.1   Tissue tropism of different AAV serotypes.

tropism	AAV serotype	publication
muscle	AAV1, AAV6-9	[104, 107, 111-113]
megakaryocytes	AAV3	[114]
retina	AAV1, AAV4, AAV5	[115, 116]
apical airway cells	AAV5, AAV6	[117, 118]
central nervous system	AAV2, AAV4, AAV5	[119]
heart	AAV9	[120]
liver	AAV8, AAV9	[104, 107, 121, 122]
kidney	AAV2, AAV9	[120, 123-125]

The differences in the tropism might be due to their binding to alternate receptors [126]. AAV2, for example, is known to bind various receptors like heparin sulfate proteoglycans (HSPG), $α_Vβ_1$ integrin, $α_Vβ_5$ integrin, fibroblast growth factor receptor (FGF-R), and hepatocyte growth factor receptor (c-met) [127-130]. The attachment to the HSPG and FGF receptor is also reported for AAV3 [131, 132], whereas AAV5 preferentially bind to sialic acid and plateled derived growth factor receptor (PDGF-R) [133, 134]. A common receptor for AAV2, 3, 8, and 9 is the laminin receptor (LamR) [126]. However, not only the binding to different cell surface receptors seems to determine vector tropism but also the intracellular processes like trafficking, endosomal escape, nuclear delivery and second strand synthesis [101].

1.5.5 Pseudo-packaging

To improve the AAV transduction efficiency for specific tissues, there is the possibility of the so called transcapsidation [135]. This process implies the feasibility of pseudotyping recombinant AAV genomes, typically derived from AAV serotype 2, with capsids from any of the other serotypes [126]. Even though AAV2 shows low transduction efficiency, the AAV tropism can be altered by packaging recombinant AAV2 genomes into capsids derived from other AAV serotypes, yielding hybrid vectors. The advantage is a modification of the targeting to cell types that express other receptors than AAV2. Several groups already cross-packaged transgenes with

AAV2 ITRs into other serotype capsids. Hildinger and colleagues, for example, cross-packaged an AAV2 ITR genome containing the *lacZ* gene into three different AAV capsids (AAV1, 2, and 5), respectively, and compared the transduction efficiency in the muscle [136], yielding improved transduction efficiency with the transcapsidated vectors compared to the wildtype. A similar approach was performed by another group with GFP as reporter, but for the retina [102].

1.5.6 Gene delivery to the kidney

Gene delivery to the kidney *in vivo* has already been shown by different gene vehicles. A non-viral gene transfer to the kidney was e.g. achieved by intrarenal-pelvic or intrarenal-arterial injections of liposomes [137], and with oligonucleotides that were intravenously injected [138]. Renal gene transfer using viral vectors are shown by Bosch and colleagues who revealed a successful gene delivery to the rat kidney via retroviral vectors [139]. In addition, an adenoviral vector mediated gene delivery into the kidney is reported by Mouillier et al. [140]. But there are also studies for the adeno-associated virus and its capability to infect the kidney. Lipkowitz and colleagues could show that an intraparenchymal injection of AAV successfully transduced renal tubular epithelial cells [141]. Also local delivery of AAV2 via the renal artery was reported to efficiently transduce tubular epithelial cells, while neither of the serotypes rAAV-1, -3, -4, or -5 showed any transduction [123]. There are also recent data of systemic transduction of the kidney by AAV9. Although this serotype is preferentially known to be a suitable vector for cardiac transduction [120], Bostick et al. reported an efficient transduction of the kidney of adult mice by AAV9 [125]. Likewise the group of Nakai demonstrated a transduction of the kidney by rAAV9 following systemic administration [142].

1.5.7 Clinical trials/therapeutical approaches of AAV

The first gene therapeutic clinical trial using a viral vector was performed 1989. Since then, at least 1471 more studies have been initiated worldwide (http://www.wiley.co.uk/genmed/clinical/). Most of the trials are addressed to cancer diseases (65.2 %), followed by cardiovascular (9.3 %), monogenic (8.2 %), and infectious diseases (7.6 %). The most commonly used viral vectors are adeno- (24.9

%) and retroviruses (21.7 %), while until now AAV vectors are used in only 4.1 % of the clinical approaches, mainly for the treatment of inherited disease and cancer. For example clinical trials using AAV as vector are reported for the treatment of the monogenic diseases cystic fibrosis and hemophilia B. For cystic fibrosis the administration of CFTR (cystic fibrosis transmembrane conductance regulator) as transgene in the nasal sinus and bronchial epithelium showed improved pulmonary function as well as partial correction of hyperinflammatory responses and electrophysical defects [143-145]. The clinical trials for hemophilia B were carried out by intramuscular [146, 147] or intrahepatic [148] administration of the vector. The muscle-directed study revealed evidence for transduction in all patients, however long-term expression of the therapeutic gene, coagulation factor IX (FIX) could only be detected at low levels. Intrahepatic administration of high vector amounts resulted in therapeutic, but transient (< 8 weeks) transgene expression levels. As no vector related adverse events were reported, this vector system proved to exhibit an excellent safety profile.

1.6 Aim of the study

The aim of this study is to analyze the anti-fibrotic effect of HGF *in vitro* and *in vivo* with regard to renal interstitial fibrosis. Still, little is known about the underlying molecular and cellular mechanisms of the anti-fibrotic actions of HGF. Therefore, the first part of this study focusses on the signal transduction of HGF and the molecular implications of the HGF-initiated signal cascades in renal interstitial fibroblasts. Furthermore, candidates selected by their pronounced divergent expression profile and their potential association with pro-fibrotic processes will be subjected to detailed analyses of the underlying regulatory mechanisms.

A further question that will be addressed in the present study is the anti-fibrotic effect of HGF *in vivo*. The aim is to establish a gene therapeutical system for the treatment of renal interstitial fibrosis, using HGF as transgene and the adeno-associated viral vector (AAV) as gene vehicle. Therefore, three different AAV serotypes shall be analyzed due to their ability to transfer efficiently HGF to the kidney. For this purpose, a promoter construct has to be generated that mediates an efficient expression of HGF in the kidney and finally, the anti-fibrotic effect of HGF has to be determined in the COL4A3 knockout mouse model that closely simulates the Alport syndrome, a disease that results in interstitial fibrosis. Fibrosis will then be evaluated and the therapeutic effect on fibrotic markers be determined.

Abbreviations

SMA	alpha smooth muscle actin
ACE	angiotensin-converting enzyme
bp	basepairs
°C	degrees celsius
cDNA	complementary DNA
DAB	3,3'- Diaminobenzidine
DABKO	(1,4-Diazabicyclo 2,2,2) octan)
DAPI	4, 6-diamidino-2-phenylindole
dd	double distilled
DMEM	Dulbecco's Modified Eagle Medium
DMSO	dimethyl sulfoxide
DNA	desoxyribonucleic acid
EDTA	ethylenediaminetetraacetic acid
ELISA	enzyme-linked immunosorbent assay
FCS	fetal calf serum
FITC	fluorescein-5-isothiocyanat
g	gram
GER	germany
GFP	green fluorescence protein
h	hour
HE	hematoxyline eosin
hHGF	human hepatocyte growth factor
H_2O	water
HRP	horseradish peroxidase
KCl	potassium chloride

ABBREVIATIONS

kDa	kilodalton
LB	Luria Bertani
mg	milligram
$MgCl_2$	magnesium chloride
$MgSO_4$	magnesium sulfate
ml	milliliter
mM	millimolar
µg	microgram
µl	microliter
µM	micromolar
µm	micrometer
NaCl	sodium chloride
NaH_2PO_4	sodium dihydrogen phosphate
Na_2HPO_4	disodium hydrogen phosphate
ng	nanogram
nm	nanometer
nt	nucleotides
NTC	no-template control
OD	optical density
PBS	phosphate buffered saline
PEG	polyethylene glycol
PCR	polymerase chain reaction
RNA	ribonucleic acid
rpm	revolutions per minute
RT	room temperature
SDS	sodium dodecyl sulfate

ABBREVIATIONS

SOB   super optimal broth
TAE   Tris acetate EDTA buffer
TBS   Tris buffered saline
TBST   Tris buffered saline tween-20

2. Materials and methods

2.1 Materials

2.1.1 Chemicals, plastic ware and other materials

Chemicals and solutions were purchased from Roth (Karlsruhe, GER), Sigma-Aldrich (Taufkirchen, GER) and Merck (Darmstadt, GER) in *pro analysi* quality if not described elsewise. Pipet tips, 0.5, 1.5, or 2 ml tubes were obtained from Biozym (Oldendorf, GER), Labomedic (Bonn, GER), or Eppendorf (Hamburg, GER). The plastic ware was autoclaved at 121 °C and 1.2 bar for 20 minutes (Varioklav, H+P Labortechnik). Glass ware was sterilized with dry heat by baking at 180 °C for 5 hours. All plastic ware used in cell culture was purchased from TPP (Hörstel, GER), Nunc (Wiesbaden, GER) and Becton-Dickinson (Lincoln Park, NJ, USA).

2.1.2 Software

The following software was used in this thesis: DNASIS®MAX (Hitachi Software Engineering Co., Ltd, Japan), Redasoft Plasmid 1.1 (REBASE Version 908), Premier Biosoft (http://www.premierbiosoft.com/netprimer/netprlaunch/netprlaunch.html), Olympus Soft Imaging System (Olympus, Hamburg, GER), Geldock, Stratagene MxPro 3000P V4.00 (La Jolla, USA), Ascent Software 2.6 (Thermo Scientific, GER), and BioRad IQ5 (BioRad, München, GER).

2.1.3 Enzymes and antibodies

2.1.3.1 Enzymes

Benzonase	(Merck Chemicals Ltd., Darmstadt, GER)
DNAse (deoxyribonuclease)	(Macherey & Nagel, Düren, GER)
Pfu Ultra™ High-Fidelity	(Stratagene, Waldbronn, GER)
Proteinase k [20mg/ml]	(Fermentas, St. Leon-Rot, GER)
Phusion-DNA-polymerase	(Biozym, Hess. Oldendorf, GER)
REDTaq®-DNA polymerase	(Sigma-Aldrich, Taufkirchen)
Restriction endonucleases	(NEB, Frankfurt, GER)
RNAse A	(Macherey & Nagel, Düren, GER)
Vent® polymerase	(Sigma-Aldrich, Taufkirchen, GER)

2.1.3.2 Antibodies

The antibodies used for Western blotting, immunohisto- or immunocytochemistry are listed in table 1 (primary antibodies) and table 2 (secondary antibodies).

Table 2.1: Primary antibodies.

antibody	raised in	manufacturer
beta-actin	mouse	Sigma, Taufkirchen, GER
collagen Iαl	rabbit	Abcam,, Cambridge, UK
phospho-Akt (Thr308)	rabbit	Cell Signaling, Frankfurt, GER
phospho-p44/42MAPK (Thr202/Tyr204)	rabbit	Cell Signaling, Frankfurt, GER
phospho-Smad2 (Ser245/250/255)	rabbit	Cell Signaling, Frankfurt, GER
phospho-Smad2 (Ser465/467)	rabbit	Cell Signaling, Frankfurt, GER
phospo-Stat3	rabbit	Cell Signaling, Frankfurt, GER
GFP	rabbit	Abcam, Cambridge, UK

Table 2.2: Secondary antibodies.

antibody	raised in	against	manufacturer
alkaline phosphatase labeled	goat	rabbit	Santa Cruz Biotechnology, Heidelberg, GER
peroxidase labeled	goat	rabbit	Dianova, Hamburg, GER
peroxidase labeled	rabbit	mouse	Dako, Hamburg, GER

2.1.3.3 Consumables

26-gauge needle	BD, Heidelberg, GER
Photographic film	Amersham Biosciences, Freiburg, GER
ReadyGel (SDS-polyacrylamide gel)	BioRad, München, GER
UV-Plastic cuvette	Eppendorf, Hamburg, GER

2.1.3.4 Devices

Microscope Eclipse TE 300	Nikon, Düsseldorf, GER
MJ Research PTC-200 Peltier PCR Cycler	GMI, Minnesota, USA
Multiscan Ascent (photometer)	Thermo Scientific, Bonn, GER
Olympus Vanos-S AH2	Olympus, Hamburg, GER
Precellys 24 Homogenisator	Peqlab, Erlangen, GER
Real-time PCR Cycler Mx3000P	Stratagene, Waldbronn, GER
Eppendorf BioPhotometer	Eppendorf, Hamburg, GER
Water bath	Dr. Hirtz & Co, Cologne, GER

2.1.3.5 Cell culture

Dulbecco's Modified Eagle Medium	Sigma-Aldrich, Taufkirchen, GER
FCS	Sigma-Aldrich, Taufkirchen, GER
Trypan blue solution (0.5 %)	Biochrom AG, Berlin, GER
Trypsin	GibcoBRL, Karlsruhe, GER

2.1.3.6 Reagents

Developer	Kodak, Stuttgart, GER
Fixing solution	Kodak, Stuttgart, GER
NuPAGE Reducing Agent (10 x)	Invitrogen, Karlsruhe, GER
NuPAGE MOPS SDS Running Buffer	Invitrogen, Karlsruhe, GER
NuPAGE LDS Sample Buffer (4 x)	Invitrogen, Karlsruhe, GER
NuPAGE Transfer Buffer	Invitrogen, Karlsruhe, GER

2.1.3.7 Cytokines

Table 2.3: Cytokines used in this study.

cytokine	species	source
recombinant HGF	human	Dianova, Hamburg, GER

2.1.4 Kits and assays

All the kits used in the experiments and the manufacturing companies are indicated below:

BCA Protein Assay	PerbioScience, Bonn, GER
BigDye terminator sequencing kit v.3.1	Applied Biosystems, Darmstadt, GER
Chemi-luminescence substrate CDP Star	Invitrogen, Karlsruhe, GER
DNeasy blood and tissue kit	Qiagen, Hilden, GER
Dual-Glo™ luciferase reporter assay	Promega, Mannheim, GER
Fast-Plasmid mini kit	Eppendorf, Hamburg, GER
High capacity cDNA RT kit	Applied Biosystems, Darmstadt, GER
human HGF Quantikine ELISA kit	R&D, Wiesbaden-Nordenstadt, GER
Lipofectamine 2000	Invitrogen, Karlsruhe, GER
Nucleobond PC-100	Macherey & Nagel, Düren, GER
Nucleospin RNA II	Macherey & Nagel, Düren, GER
Nucleospin RNA/Protein	Macherey & Nagel, Düren, GER
Perfectprep Gel Cleanup kit	Eppendorf, Hamburg, GER
Plasmid-Mega-kit	Qiagen, Hilden, GER
Puregene DNA isolation kit	Biozym, Hess. Oldendorf, GER

2.1.5 Oligonucleotides

All oligonucleotides used in this thesis were ordered from Eurofins MWG Operon (Ebersberg, GER) and are listed in Table 2.4.

MATERIALS AND METHODS

Table 2.4: Oligonucleotides

oligonucleotide	species*	sequence	length (nt)
Ksp-Cad-F	m	TAT ACT CGA GAG CTT GCT CTG CCA TG	26
Ksp-Cad-R	m	TAT TAA GCT TCT TCA GGG AGC TCT GGC	27
Ksp-Cad-Seq1	m	CTA GGC TTC TGT CCC ACC CAC	21
Ksp-Cad-Seq2	m	GTA TTA GCT TCG GAG TTC CTC TG	23
Ksp-Cad-Seq3	m	GGC ACA AGG AAC AA TAT CTG	20
Ksp-Cad-R-Mut	m	CAA GTG GCC CGT GGA GCT AAG G	22
ß-globin-min-F	m	AGC TTC TGG GCA TAA AAG TCA GGG CAG AGC CAT CTA TTG CTT ACA TTT GCT TCT GGA ATT CA	62
ß-globin-min-R	m	AGA CCC GTA TTT TCA GTC CCG TCT CGG TAG ATA ACG AAT GTA AACtGAA GAC CTT AAG TTC GA	62
CMV-enhancer-F	m	ATT ACG GGG TCA TTA GTT CAT AGC	24
CMV-enhancer-R	m	ACA TTT TGG AAA GTC CCG TTG	21
hHGF-F	h	TCT AAT AAG CTT GCC AAC ATG TGG GTG ACC AAA CTC	36
hHGF-R	h	CTC TAC GTC GAC CTG GCC TTT TGC TCA CAT GTT C	34
COL4A3-1	m	CCA GGC TTA AAG GGA AAT CC	20
COL4A3-2	m	TCT GCT AAT ATA GGG TTC GAG A	22
COL4A3-3	m	GCT ATC AGG ACA TAG CGT TGG	21
Actin-F	r	CTA GAC TTC GAG CAG GAG ATG GC	23
Actin-R	r	GAA TGT AGT TTC ATG GAT GCC AC	23
mCRP-F	m	ACC CAC ATT GAT TTC TCT GTT CTA	24
mCRP-R	m	AAT GAT TTC CTA ACA CTG CCT CTT	24
COL1A1-F	m	CAT GTT CAG CTT TGT GGA CCT	21
COL1A1-R	m	GCA GCT GAC TTC AGG GAT GT	20
HPRT-F	m	TCC TCC TCA GAC CGC TTT T	19
HPRT-R	m	CCT GGT TCA TCA TCG CTA ATC	21
CTGF-F	r	GCT GAC CTA GAG GAA AAC ATT AAG A	25
CTGF-R	r	CCG GTA GGT CTT CAC ACT GG	20
CTGF-F	m	TGA CCT GGA GGA AAA CAT TAA GA	23

MATERIALS AND METHODS

CTGF-R	m	AGC CCT GTA TGT CTT CAC ACT G	22
HPRT-F	r	GAC CGG TTC TGT CAT GTC G	19
HPRT-R	r	ACC TGG TTC ATC ATC ACT AAT CAC	24
mCRP-F	m	ACC CAC ATT GAT TTC TCT GTT CTA	24
mCRP-R	m	AAT GAT TTC CTA ACA CTG CCT CTT	24
mCRP probe	m	TCC CTT TCT CCC AGT GGT CTG ACG T	25
COL1A1-F	r	CAT GTT CAG CTT TGT GGA CCT	21
COL1A1-R	r	GCA GCT GAC TTC AGG GAT GT	20
TGFβ-F	r	CCT GGA AAG GGC TCA ACA C	19
TGFβ-R	r	CAG TTC TTC TCT GTG GAG CTG	22
α-SMA-F	m	ACT CTC TTC CAG CCA TCT TTC A	22
α-SMA-R	m	ATA GGT GGT TTC GTG GAT GC	20
α-SMA-F	r	TGC CAT GTA TGT GGC TAT TCA	21
α-SMA-R	r	ACC AGT TGT ACG TCC AGA AGC	21
Smad2-F	r	CAG GAC GAT TAG ATG AGC TTG A	22
Smad2-R	r	CCC CAA ATT TCA GAG CAA GT	20
Smad3-F	r	CCT GCC ACT GTC TGC AAG	18
Smad3-R	r	GCA GCA AAT CCT GGT GG TT	20
Smad4-F	r	GAA CAC TGG ATG GAC GAC TG	20
Smad4-R	r	ACA GAC GGG CAT AGA TCA CA	20
Wisp-2-F	m,r	CAG GGC CTG GTT TGT CAG	18
Wisp-2-R	m,r	AGC TAC CGT CAT CCT CAT CC	20
FIGF-F	m,r	TGT TTT ACA AGA TGA GAA TCC ACT G	25
FIGF-R	m,r	GGG TTC CTG GAG GTA AGA GTG	21
Nov-F	m,r	ATG GTT CGG CCT TGT GAG	18
Nov-R	m,r	TTG GTC CGG AGA CAC TTT TT	20
PDGFRβ-F	m,r	TCT CTC ATC ATC CTC ATC ATG C	22
PDGFRβ-R	m,r	CCT TCC ATC GGA TCT CAT AGC	21
Bambi-F	m,r	TCA TCT GGC TGC AGT TGG	18
Bambi-R	m,r	CAT CAC AGT AGC ATC TGA TCT CG	23
GFP-F		ATG GTG AGC AAG GGC GAG GA	20
GFP-R		GGA CAC GCT GAA CTT GTG GC	20
GFP-probe		TTA CGT CGC CGT CCA GCT CGA CCA G	25

* m = mouse, r = rat h = human

2.1.6 Plasmids

Table 2.5: Plasmids used in this thesis.

vector	specification/characteristic	source/reference
pGL3-Basic	Firefly luciferase vector	Promega
pGL3-Ksp	Ksp-cadherin promoter, firefly luciferase	this thesis
pGL3-CMV-Ksp	CMV-Enhancer, Ksp-cadherin, firefly luciferase	this thesis
pGL3-Ksp-β-globin	Ksp-cadherin promoter, β-globin minimal promoter, firefly luciferase	this thesis
pGL3-CMV-Ksp-hHGF	CMV-Enhancer, Ksp-cadherin promoter, hHGF	this thesis
pSUB201-plus	AAV-packaging plasmid	AG Büning
pSUB-CMV-Ksp-hHGF	CMV-Enhancer, Ksp-cadherin promoter, hHGF flanked by ITR's	this thesis
pEGFP-C1	CMV promoter, EGFP	Clontech
psc-GFP	CMV promoter, EGFP (self complementary)	AG Büning
pXX6	adenoviral genes (E2A, VA, E4)	J. Samulski, University of North Carolina
pRC	Rep (AAV2), Cap (AAV2)	A. Girod, Universität München
pXR8	Rep (AAV2), Cap (AAV8)	J. Wilson, University of Pensylvania
pXR9	Rep (AAV2), Cap (AAV9)	J. Wilson, University of Pensylvania

2.1.7 Buffers and solutions

All buffers and solutions were prepared using completely desalted millipore purified water (Millipore-Q Plus, Millipore, (Molsheim, GER)).

LB medium:	1.0 % [w/v]	tryptone (Fluka, GER)
	0.5 % [w/v]	bacto-yeast extract (Difco, USA)
	0.8 % [w/v]	NaCl
	NaOH	(adjust pH value to 7.6)
SOB medium:	2.0 % [w/v]	tryptone
	0.5 % [w/v]	bacto-yeast extract
	10.0 mM	NaCl
	2.5 mM	KCl
	10.0 mM	$MgSO_4$
	20.0 mM	glucose
LB agar:	15 g /l	LB medium Bacto- Agar
2 x TSS:	20 % [w/v]	PEG 8000 (Sigma, GER)
	10 % [v/v]	DMSO (Sigma, GER)
	70.0 mM	$MgCl_2$ in LB
	pH 6,5	
Lysis buffer for DNA extraction	50 mM	TRIS-HCl, pH 8.0
	100 mM	EDTA
	0.125 % [w/v]	SDS
	0.8 mg/ml	proteinase K (Fermentas, GER)
Phosphate-buffered saline (PBS) 10 x	74 g	NaCl
	14.2 g	$Na_2HPO_4 \times H_2O$
	3.62 g	$NaH_2PO_4 \times 2H_2O$
	ad 1000 ml	dH_2O
	pH	7.4
Tris-buffered saline (TBS) 10 x	150 mM	NaCl
	20 mM	TRIS-HCl
	ad 1000 ml	dH_2O
	pH	7.6

SDS-protein lysis buffer	15 mM	TRIS-HCl (pH 6.8)
	2.5 % [v/v]	glycerol
	0.5 % [w/v]	SDS
	1 mM	complete (Roche, GER)

NuPAGE SDS sample buffer (4 x) Invitrogen, Karlsruhe, GER

NuPAGE SDS running buffer (20 x) Invitrogen, Karlsruhe, GER

NuPAGE SDS transfer buffer (20 x) Invitrogen, Karlsruhe, GER

Ponceau staining solution	0.5 % [w/v]	Ponceau S
	1 % [v/v]	acetic acid
Destaining solution	10 % [v/v]	methanol
	10 % [v/v]	acetic acid
TAE buffer	40 mM	Tris-acetate, pH 7.8
	5 mM	Na-acetate
	1 mM	EDTA, pH 8.0

2.1.8 Cell lines

Cell lines are shown in table 2.6.

Table 2.6: Cell lines.

cell line	origin	characteristics	reference	source
HUH7	human	hepato cellular carcinoma cell line	LGC Promochem	P. Schirmacher
HeLa	human	Human cervix carcinoma cell line	DSMZ-No.: ACC 57	DSMZ
HEK293	human	embryonic kidney cell line	DSMZ-No.: ACC 305	DSMZ
NRK52E	rat	kidney epithelial-like cells	DSMZ-No.: ACC 199	DSMZ
NRK49F	rat	kidney fibroblasts	DSMZ-No.: ACC 172	DSMZ

2.2 Methods

2.2.1 DNA preparation

2.2.1.1 DNA preparation of *E. coli*

All centrifugation steps of the procedure were performed in an Eppendorf centrifuge Type 5417R (Eppendorf, Hamburg, GER). 5 ml LB medium were inoculated and the *E. coli* cultures were grown overnight. 1.5 ml of the bacteria culture were centrifuged at 14000 rpm for 1 min, cells were then resuspended in 150 µl buffer S1 (Midiprep kit Nucleobond® 100, Macherey & Nagel, GER) and the DNA was extracted according to the manufacturer's instructions. Briefly, 150 µl buffer S2 were added. After incubation for 5 min at RT, lysis was stopped via addition of 150 µl S3 and incubation on ice for 5 min. To remove denatured proteins and cell debris, a further centrifugation at 14000 rpm for 5 min was done and the plasmid DNA in the supernatant was precipitated by addition of 500 µl 100 % isopropanol and centrifugation at 14000 rpm for 5 min. The pellet was washed with 70 % ethanol, air dried and re-dissolved in 20 µl deionized water.

For the preparation of a larger quantity of plasmid DNA, a 100 ml or 500 ml culture was inoculated and grown overnight. The plasmid DNA was extracted by using the Nucleobond PC-100 kit (Macherey & Nagel, GER) or the Plasmid-Mega-kit (Qiagen, GER), respectively, according to manufacturer's instructions. The eluted, purified plasmid DNA was precipitated with isopropanol, washed with 70 % ethanol, air dried and re-dissolved in deionized water.

2.2.1.2 DNA extraction by phenol chloroform extraction

All centrifugation steps of the procedure were performed in an Eppendorf centrifuge Type 5417R (Eppendorf, GER).
To purify DNA from a solution that also contains proteins, the DNA was extracted by standard phenol chloroform extraction. An equal volume of phenol-chloroform-isoamylalcohol (Roth, GER) was added to proteinase k-lysed cells or tissue, well mixed and centrifuged at 14000 rpm for 5 min. The aqueous top layer was transferred to a new tube and extracted for a second time with an equal volume of

phenol-chloroform-isoamylalcohol. To avoid phenol contamination of the samples, the aqueous top layer was extracted twice with an equal volume of chloroform-isoamylalcohol. Finally, the aqueous top layer was transferred to a new tube and precipitated with 1:10 vol. 3M sodium acetate pH 5.2 and 2 vol. ice cold 100 % ethanol. The pellet was washed with 70% ethanol, air-dried and re-dissolved in an appropriate volume of deionized water.

2.2.1.3   DNA extraction from mouse tails

The genomic DNA extraction from mouse tail was performed according to manufacturer's instructions using the PureGene DNA isolation kit (Biozym, GER).

2.2.2 RNA preparation

2.2.2.1   RNA isolation

The isolation of total RNA from cells or tissues was performed using the Nucleospin total RNA II kit (Macherey & Nagel, GER). The isolation was carried out according to the manufacturer's instructions. After cell lysis and filtration of the lysate, the RNA was bound to a silica membrane. The DNA was digested by a DNase step and the RNA was eluted with 60 µl RNase-free water and stored at -70 °C until further usage.

2.2.3 DNA modification

2.2.3.1   Restriction analysis

All restriction enzymes were purchased from NEB (Frankfurt, GER), Fermentas (St. Leon- Rot, GER) and Roche (Mannheim, GER). Cleavage of DNA with restriction enzymes was done using standard methods according to the instructions of the manufacturer. For a double digestion, if no common buffer specific to both enzymes was available, the plasmid DNA was ethanol-precipitated following the first digestion and re-dissolved in the appropriate buffer of the second enzyme.

2.2.3.2 Dephosphorylation by alkaline phosphatase

Alkaline phosphatase catalyzes the removal of the 5' phosphate groups from DNA, RNA and ribo- and deoxyribonucleotide triphosphates. DNA was dephosphorylated using the Shrimp alkaline phosphatase (SAP, Roche, GER) according to the instructions of the manufacturer. After the reaction, the SAP was heat-inactivated (15 min, 65 °C).

2.2.3.4 Generation of blunt ends

For the removal of 3' overhang of double-strand DNA, the T4 polymerase (Fermentas, GER) with a 3'→5' exonuclease activity was used. The reaction was incubated at 11 °C for 30 min and then heat-inactivated for 10 min at 75 °C.

2.2.3.5 Dimerisation and phosphorylation of oligonucleotides

Dimerisation of complementary oligonucleotides was initiated by heating the dimerisation mix to 95 °C for 10 min followed by slowly cooling down. The dimerisation was done according to the following protocol:

Dimerisation mix:	oligo-nucleotide 1 [100 µM]	20 µl
	oligo-nucleotide 2 [100 µM]	20 µl
	hybridization solution	40 µl
hybridisation solution:	TRIS	50 mM
	NaCl	300 mM
	EDTA	2 mM

Phosphorylation of the 5´-ends of the dimers was carried out using the T4-Polynucleotidkinase (PNK) of Fermentas (St. Leon-Rot, GER).

Phosphorylation mix: oligo-nucleotide dimer [25 µM]   1.0 µl
                              ATP [625 mM]   0.4 µl
                              T4 DNA polymerase buffer   1.0 µl

2.2.3.6 Ligation of DNA fragments

The ligation reactions using T4 DNA ligase (Fermentas, GER) were prepared according to the manufacturer's instructions and incubated at room temperature for 2 hours or at 14.5 °C overnight.

2.2.4 Transformation

The ligation reaction (2.3.5) was added to 100 µl of competent cells (DH5α) and incubated on ice for 30 min. Transformation efficiency was enhanced by a subsequent heat shock (2 min, 42 °C), followed by chilling bacteria on ice for a further 5 min. For regeneration and development of antibiotic resistance, the bacteria were incubated with 600 µl SOC-medium at 37 °C for 45-60 min. The cells were plated on LB-agar/ampicillin (100 µg/ml) or LB-agar/kanamycin (100 µg/ml) plates, inverted and incubated overnight at 37 °C.

2.2.4.1 Preparation of competent bacteria

Competent bacteria were prepared by the TSS method described by Chung et al. (1989). 100 ml LB medium were inoculated with 200 µl of a *E. coli* DH5α overnight pre-culture and cultivated until the early log-phase (OD_{600} = 0.4). Cells were harvested by centrifugation (5000 rpm for 15 min) and resuspended in 5 ml LB medium. An equal volume of TSS (2x) was added, carefully mixed and directly shock-frozen as 150 µl aliquots in liquid nitrogen. Cells were stored at -70 °C until further usage.

TSS (2x):   20 %   PEG8000
            10 %   DMSO
            70 mM   MgCl2
            dissolve in LB and adjust pH to 6.5

2.2.5 Polymerase chain reaction (PCR)

The polymerase chain reaction is a method to amplify a certain region of DNA in an exponential manner using a thermostable DNA polymerase. The DNA polymerase normally used in this reaction, originates from the bacterium *Thermus aquaticus*. The heat resistance enables the polymerase to endure the steps of high temperature during the PCR. Furthermore, two primers are required for the PCR. The nucleotides for the newly synthesized DNA strands are supplied as nucleotide triphosphates. A PCR is normally characterized by three steps of different temperatures (denaturation, annealing and elongation). These three steps are repeated several times resulting in an exponential amplification of the target DNA strand. The exact temperatures for a PCR depend on the template as well as the composition of the primers. Thus, the conditions were different in each PCR application.

2.2.5.1 Qualitative PCR

For the detection of the presence of a defined DNA sequence in a sample or for preparative approaches, qualitative PCR was used. PCR was performed in a total volume of 25 µl. The components of the reaction (dNTP's, buffer, primer, polymerase and template) were applied to the mix according to the manufacturer's instructions. The Taq-DNA-polymerases that were used in the assays are listed in table 2.7.

Table 2.7: DNA polymerases.

DNA polymerase	proof-reading activity	company
Triple-Master™	3'→5'	Eppendorf, GER
Pfu Ultra™ High-Fidelity	3'→5'	Stratagene, GER
Vent®	3'→5'	NEB, GER
REDTaq®	--	Sigma, GER

The PCR were performed according to the manufacturer's instructions unless otherwise noted. An example for cycling conditions is:

1.	94 °C	5 min	
2.	94 °C	1 min	denaturation
3.	54 – 65 °C	30 sec	annealing
4.	72 °C	30 sec	elongation
5.	72 °C	10 min	

The steps 2 to 4 were repeated 34 times.

2.2.5.2 Genotyping of mice

Genotyping of mice was performed by PCR. The PCR was carried out on 0.5 µl tail lysate with REDTaq polymerase (Sigma, GER) in a reaction mixture as follows:

PCR reaction mixture:

DNA:	0.5 µl
COL4A3-1:	2 µl
COL4A3-2:	2 µl
COL4A3-3:	0.25 µl
DMSO:	1 µl
Red-Taq:	12.5 µl
water:	ad 25 µl

PCR conditions for genotyping of COL4A3 mice were: 94 °C for 2 min; 45 cycles of 94 °C for 30 sec, 56 °C for 60 sec, 72 °C for 2 min; and 72 °C for 2 min.

The resulting PCR fragments supplied information about the genotype: a single fragment of 900 bp indicated a wildtype mouse, two fragments of 280 bp and 900 bp a heterozygous one, and two fragments (280 bp+1900 bp) a knockout mouse.

2.2.5.3 Real-time PCR

Real-time PCR is a quantitative PCR method for the determination of the copy number of PCR templates such as DNA or cDNA in a PCR. It monitors the increase of DNA as it is amplified via fluorescence emitted during the reaction as an indicator. There are two types of real-time PCR: probe-based and intercalator-based. Probe-based real-time PCR requires, in addition to PCR primers, a fluorochrom-labeled probe which is an oligonucleotide with both, a fluorescent reporter at one end, and a quencher of fluorescence, at the opposite end. The 5'→3' activity of the Taq polymerase breaks down the probe resulting in the breakdown of the reporter-quencher proximity allowing unquenched emission of fluorescence.

An increase in the product targeted by the reporter probe at each PCR cycle therefore causes a proportional increase in fluorescence due to the breakdown of the probe and release of the reporter. The used alternative, the intercalator-based method, is also known as SYBR Green method. It requires a dye, named SYBR Green, in the PCR which binds to newly synthesized double-stranded DNA and gives fluorescence, thus determining the amplicon production. SYBR Green is a minor groove binding dye that does not bind to ssDNA and whose fluorescence is greatly enhanced by binding. During the stages of PCR, different intensities of fluorescence signals can be detected, depending on the amount of dsDNA that is present.

For the reaction with unspecific dyes, 10 µl SyBr Green I PCR Master Mix (ABI, Darmstadt, GER) were mixed with 0.8 µl forward and reverse primer (both 10 µM), 7.4 µl deionized H_2O and 1 µl DNA or cDNA (10-50 ng). The standard curve was made either by a DNA dilution series (10^8 to 10^1 molecules) of a plasmid containing the coding sequence of interest or by a cDNA dilution series (50 ng – 3.12 ng). The PCR was performed at the following conditions. The steps 2 and 3 were repeated in 55 cycles:

1.	95 °C	10 min	
2.	95 °C	30 sec	
3.	60 °C	1 min	
4.	95 °C	1 min	
5.	55 °C	30 sec	dissociation curve
6.	95 °C	30 sec	

The GFP probe (Sequence: TTACGTCGCCGTCCAGCTCGACCAG) was used as specific probe. 8 µl RealMasterMix Probe were mixed with 0.6 µl forward and reverse primer (both 10 µM), 0.2 µl probe (10 µM), 9.6 µl deionized H_2O and 1 µl DNA or cDNA (50 ng). A dilution series (10^8 to 10^2 molecules) of a plasmid containing the GFP coding sequence was used as standard curve. The PCR was performed at the following conditions.

The steps 2 and 3 were repeated in 55 cycles:

1.	95 °C	2 min
2.	95 °C	30 sec
3.	60 °C	1 min

For both methods the real-time PCR was run on a Mx3000P Cycler (Stratagene, GER). In order to determine the number of copies that are contained in a sample, a standard curve was prepared by plotting the copy number versus the Cycle Threshold.

2.2.5.4 Determination of the titer of AAV preparations or the AAV infection rate by quantitative PCR

In order to evaluate the titer rate of virus preparations or the infection rate of AAV, a dilution series of the respective mother plasmid was used as a calibration series with defined copy numbers.

2.2.5.5   Sequencing of DNA

The sequencing reactions were all performed using the chain termination method developed by Frederick Sanger (Sanger et al., 1977). The sequencing was done according to the "cycle sequencing" technique using the BigDye terminator sequencing kit v.3.1 (Applied Biosystems, GER). The resulting DNA fragments of varying length were separated by capillary gel electrophoresis and analysed by an automated sequencer (ABI 3730) in the service laboratory of the Institute of Genetics (University of Cologne).

Sequencing mix:
BigDye terminator mix   2 µl
buffer   4 µl
primer   5 pmol
DNA   150 – 300 ng
H_2O ad 20 µl

Cycling:

PCR conditions for sequencing were:

1. 96 °C   1 min
2. 96 °C   15 sec
3. 50 °C   10 sec
4. 60 °C   4 min
5. 60 °C   10 min

The steps 2 to 4 were repeated 30 times.

2.2.5.6   Reverse transcriptase reaction

For real-time PCR, mRNA was reverse transcribed into cDNA using the High Capacity cDNA RT Kit. As primers, random hexamer oligonucleotides were added. Annealing of the random primers took place at 25 °C for 10 minutes followed by 2 hours cDNA synthesis at 37 °C. The cDNA reaction mixture was according to manufacturer's instructions.

2.2.6 Biochemical methods

2.2.6.1 Preparation of protein extracts and determination of protein concentration

Cell monolayers were rinsed twice with 1x PBS and scraped in SDS protein lysis buffer. After three freeze and thaw cycles, the protein extracts were centrifuged for 15 min at 14.000 rpm at 4 °C in an Eppendorf centrifuge Type 5417R (Eppendorf, Hamburg, GER). Protein concentrations of the supernatants ("protein extracts") were determined using the BCA Protein Assay (PerbioScience, Bonn, GER) with a dilution series of BSA for calibration according to the recommendations of the supplier. Colour change of the samples were read out at 570 nm wavelength in a Multiscan Ascent photometer (Thermo Scientific, Bonn, GER).

2.2.6.2 Western blot analysis

SDS-polyacrylamide gel electrophoresis (SDS-PAGE) was performed using the Bio-Rad Mini protein gel system. Gradient gels (4-12 %) were obtained from Bio-Rad (München, GER). Protein samples were mixed with NuPAGE SDS sample buffer (4 x), denatured at 70 °C for 10 min and directly loaded onto the gels. Electrophoresis was carried out at 100 V for about 1 h. The molecular weight of proteins was estimated by running pre-stained (See blue marker, Invitrogen, Karlsruhe, GER) marker proteins. After separation, the proteins were transferred to a nitrocellulose membrane (Amersham-Bioscience, Freiburg, GER) using the semi-dry NuPAGE-blotting system from Invitrogen for 90 min at 30 V. After transfer, the immobilized proteins were visualized by Ponceau staining (Sigma-Aldrich, Taufkirchen, GER) according manufacturer´s instructions. The membranes were then incubated in blocking solution (1 x TBS 5 % [w/v] dry milk, 0.2 % [v/v] Tween-20) for 1 h at RT. Primary antibodies were either diluted in blocking solution or 5 % BSA in 1 x TBS and 0.1 % Tween-20 (table 2.1) and incubated at 4 °C overnight with gentle shaking. Membranes were washed with TBST (10 mM TRIS-HCl, 100 mM NaCl, 0.05 % Tween, pH 7.4) four times for 15 min, each, followed by two additional washes with 1 x TBS. Membranes were then incubated for 1 h at RT with the appropriate secondary antibody alkaline phosphatase conjugates (table 2.2), diluted in blocking solution. Membranes were washed four times for 15 min with TBST and twice with 1 x TBS. Western blots were developed using chemiluminescence substrate (CDP Star, Invitrogen, GER) according to the manufacturer's instructions. Blots could be stripped by incubation in stripping buffer (1 x TBS with 2 % SDS and 7 µl β-mercaptoethanol / ml) for 30 min with gentle agitation. After extensive washing in 1 x TBS membranes could be re-probed.

2.2.7 Cell culture

2.2.7.1 Cell culture medium

HEK293, HeLa, NRK52E and NRK49F cells were grown in Dulbecco's modified Eagle's medium with 4.5 g/l glucose (DMEM) (Sigma, St. Louis, MO, USA) supplemented with heat-inactivated 10 % (v/v) fetal calf serum (FCS) (Sigma, St. Louis, MO, USA), 100 U/ml penicillin and 100 µg/ml streptomycin at 37 ºC, 5 % CO_2. Cells were regularly passaged when 80 % confluency was reached.

2.2.7.2 Passage of cells

Adherent cells were washed with 1 x PBS to remove traces of serum which inhibits trypsin. The medium was aspirated from the monolayers, 1 x PBS was added and directly aspirated again. The cells were covered with trypsin (0.25 % trypsin / 0.02 % (w/v) EDTA in 1 x PBS) and after 10 seconds, most of the trypsin was aspirated. The cells were incubated at 37 °C for 1-5 min until the adherence was abolished. Trypsin was stopped by adding fresh medium and cell clumps were broken up by pipetting vigorously. The cells were split and further cultivated in DMEM.

2.2.7.3 Cryopreservation of cells

Monolayers of 80-90 % confluency were washed with 1 x PBS, trypsinized, assimilated in fresh medium and centrifuged at 800 rpm / 4 °C for 5 min in a Beckman GPR centrifuge (Beckman, Krefeld, GER). The pellet was resuspended in 500 µl fresh medium and mixed with DMSO to a final concentration of 10 %. The cells were slowly cooled down at -70 °C over 24 h and finally stored in liquid nitrogen.

In order to culture cryopreserved cells, they were rapidly thawed and incubated in a culture dish in 9 ml fresh medium. The removal of DMSO leftovers was ensured by a medium change 12 h later.

2.2.7.4 Stimulation of cells

Cells were trypsinized and seeded in an appropriate density to gain 70-75 % confluency 12 h later. For the following 24 hours, the cells were serum starved by medium change to DMEM with 0.5 % FCS. The stimulation was done by adding the relevant cytokine (hHGF [40 ng ml^{-1}]).

2.2.7.5 Transfection of plasmid DNA

Plasmid transfections were performed using Lipofectamine 2000 (Invitrogen) according to manufacturer's instructions. Cells used for transfection were freshly passaged one day before transfection and cultivated in antibiotic-free medium. At time of transfection, the cells had a confluency of 90 %. Alternatively, the calcium precipitation method was used. This technique relies on precipitates of plasmid DNA formed by its interaction with calcium ions finally absorbed endocytotically by the cell. The cells were plated one day prior to transfection. After 24 h and a confluency of 80 %, the medium was replaced and after a further two hours the cells were transfected. For each 150-mm-diameter cell culture dish an equal molar ratio of DNA was mixed with 1 ml of 250 mM CaCl$_2$ and 1 ml HBS buffer was slowly added and briefly vortexed. The mix was incubated at RT for 2 min and then added to the cells.

Transfection buffer (HBS) 2x:	Hepes	50 mM
	NaCl	280 mM
	Na$_2$HPO$_4$	1.5 mM
	pH 7.2	

2.2.7.6 HGF ELISA

Determination of human HGF levels in cell culture medium or sera of mice were performed using the commercial Quantikine Human HGF Immunoassay of R&D (Wiesbaden, GER) according to manufacturer's instructions.

2.2.8 Functional Analysis

2.2.8.1 Luciferase Assay

The luciferase assays in the experiments were performed using the Dual-Glo™ luciferase reporter assay system of Promega (Mannheim, GER) according to manufacturer's instructions. The assay was performed on a plate reading luminometer with integrated dispenser. The co-transfected cells were washed with 1 x PBS, lysed in an appropriate volume of lysis-buffer and 20 µl of each sample was measured in duplets.

2.2.9 AAV production

2.2.9.1 Preparation of AAV

7.5×10^6 HEK293 cells (passage <30) were plated one day prior to transfection on a 150-mm-diameter cell culture dish, 30 dishes for each virus preparation. The cells were transfected by the calcium phosphate precipitation method (2.7.5). For each 150-mm-diameter cell culture dish an equal molar ratio of adenoviral helper plasmid (22.5 µg), transgene plasmid (7.5 µg) and helper plasmid (7.5 µg) were added. 24 h after transfection and incubation at 37 °C and 5 % CO_2, the medium was changed to DMEM supplemented with 2 % FCS and 100 U/ml penicillin and 100 µg/ml streptomycin and incubated for a further 24 hours. The HEK293 cells were harvested by scraping and centrifugation in a Beckman GPR centrifuge at 1200 rpm for 15 min and the cell pellet was resuspended in 7 ml lysis buffer.

MATERIALS AND METHODS

Lysis buffer: 50 mM   TRIS-HCl
150 mM   NaCl
pH 8.5

2.2.9.2   AAV extraction and purification

Virus particles were released by three freeze-and-thaw cycles. A subsequent Benzonase digestion (50 U/ml) for 30 min at 37 °C resulted in the removal of cellular DNA and RNA. The digestion was followed by a centrifugation at 3220 x g / 60 min / 4 °C and the virus-containing supernatant was purified via density gradient centrifugation.

To set up the gradient the supernatant was transferred to an ultracentrifuge tube (Thermo Scientific, Bonn, GER) and underlayed with 15 % (9 ml), 25 % (6 ml), 40 % (5 ml) and 60 % iodixanol solution (5 ml). Finally the gradient was filled up with PBS/Mg (1 mM) / KCl (2.5 mM). The centrifugation was carried out for 2 hours at 63000 rpm and 4 °C in a Sorvall Ultracentrifuge OTD Combi (Thermo Scientific, Bonn, GER). Subsequently, the 40 % iodixanol gradient layers, containing AAV vector, were harvested. The AAV preparations were stored at -70 °C and thawed on ice until further usage.

Table 2.8:   Iodixanol solutions.

	15 %	25 %	40 %	60 %
10 x PBS	5 ml	5 ml	5 ml	/
1 M $MgCl_2$	50 µl	50 µl	50 µl	50 µl
2.5 M KCl	50 µl	50 µl	50 µl	50 µl
5 M NaCl	10 ml	10 ml	/	/
Optiprep	12.5 ml	20 ml	33.3 ml	50 ml
0.5 % phenolred	75 µl	75 µl	/	25 µl
H_2O	ad 50 ml	ad 50 ml	ad 50 ml	ad 50 ml

2.2.10 The COL4A3 knockout mouse model

The mouse model used in the experiments was the COL4A3 knockout mouse [16], a mouse model of autosomal-recessive Alport syndrome. COL4A3 knockout 129/Sv mice (Jackson Immunoresearch Laboratories, Westgrove, PA, USA) were kindly provided by O. Gross [149] and had free access to regular chow and water. Only heterozygous COL4A3 knockout 129/Sv mice were crossbred. All experiments were conducted in accordance with National Health and Medical Research Committee Guidelines for Animal Experimentation.

2.2.10.1 Transduction of mice with AAV

4 week old homozygous COL4A3 mice were systemically transduced with 5×10^{11} iodixanol purified AAV particles via the tail vein and were sacrificed either 2 weeks later for reporter gene analyses or at the age of 9.5 weeks for the gene therapeutical approach. The control mice were transduced with the same quantity of AAV particles, containing empty capsids instead of a transgene.

2.2.10.2 Preparation of organs from adult mice

Adult mice were sacrificed by cervical dislocation or decapitation (in case of blood extraction for serum) and the abdomen was opened. The organs (kidney, liver, heart and spleen) were removed and shock frozen for cryosections and DNA- and RNA isolation and fixed for paraffin embedding.

2.2.10.3 Fixation, paraffin embedding and microtoming of mouse organs

Organs, fixed in 10 % phosphate-buffered formalin overnight at RT followed by 9 hours automated processing with series of dehydration steps and were embedded in paraffin wax. Paraffin blocks were sectioned with a microtome and 5 µm sections were floated on a water bath at 50 °C, took up on glass slides, dried overnight at 37 °C and stored at room temperature until staining.

2.2.10.4 Morphological and immunohistochemical studies

All stainings for histological evaluation were kindly performed by the routine laboratory of the Institute for Pathology, University Hospital Cologne.

2.2.10.5 Classification of the fibrosis grade in transversal kidney sections

For the evaluation of the fibrosis grade, transversal kidney sections were examined by gomori staining. Interstitial fibrosis, predominantly visible by ECM deposition, and the severity of fibrosis was graded as follows: grade 1: normal; grade 2: ECM accumulation in less than 5 % of the section; grade 3: ECM accumulation in less than 10 % of the section; grade 4: ECM accumulation in less than 20 % of the section; grade 5: ECM accumulation in more than 20 % of the section.

2.2.10.6 Immunohistochemical staining for α-SMA

Paraffin-embedded sections were deparaffinized by two rinses with xylene and three rinses with alcohol (100 %, 96 %, and 75 %). Endogenous peroxidase activity was quenched by an incubation of the sections in 0.3 % H_2O_2 in methanol for 30 min. Afterwards, the sections were incubated for 30 min in blocking solution (10 % milk powder in PBS 1 x) followed by an overnight incubation with the ready to use smooth muscle actin HRP conjugated primary antibody. As substrate for the peroxidase, the DAB solution, was applied to the sections and enzymatic reaction was monitored by colour development. The reaction was stopped by diving the slides in PBS. Finally the sections were counterstained with haematoxylin, washed in tap water and covered with pertex.

| **DAB solution:** | 10 mg | DAB tablet |
| | 15 ml | PBS (1 x) |

The solution was filtered and 12 µl of 0.3 % H_2O_2 were added.

2.2.10.7 Immunohistochemical staining for GFP

Paraffin-embedded sections were stained with anti-GFP antibody. After deparaffinization steps (see 2.2.10.6), sections were pre-treated in a microwave oven at 350 watt for 10 minutes in citrate buffer (pH 6.0). This treatment was followed by a blocking reaction with avidin in goat normal serum (diluted 1:100 in 10 % milk powder in PBS) for 30 min at RT going along with a biotin blocking (1:100 in 10 % milk powder in PBS) for 15 minutes at RT. The primary antibody, recognizing GFP (1:200), was incubated overnight at 4 °C. GFP antibody binding was detected using the ABC method from Vector Laboratories (Burlingame, CA, USA) following the instructions of the manufacturer. Finally, the sections were counterstained with haematoxylin. After washing for 10 minutes in tap water the slides were covered with glycerine gelatine mounting medium.

3. Results

HGF was demonstrated to act as an anti-fibrotic agent [74]. However, the molecular and cellular mechanisms underlying the anti-fibrotic activities of HGF are not well understood. In the present study anti-fibrotic signaling of HGF was analyzed and the resulting transcript profiles affected by HGF were studied. In order to apply the anti-fibrotic actions as therapeutical tool to renal fibrosis, a mouse model for renal interstitial fibrosis, the COL4A3 knockout mice, was used. To enable a most efficient delivery of HGF to the renal tubulointerstitium, different serotypes of the adeno-associated virus (AAV) were chosen as gene vehicle and analyzed.

3.1 HGF acting as an anti-fibrotic agent

An important cell type in tubulointerstitial fibrosis is represented by the renal interstitial fibroblasts. These cells play a decisive role in the accumulation of extracellular matrix and therefore the effect of hHGF on this cell type was studied by *in vitro* analyses.

To allow an accurate stimulation of renal fibroblasts by hHGF, the cell line NRK49F used in this study was examined with regard to the expression of the HGF receptor c-met. Real-time PCR using c-met specific primers designed to hybridise to the c-met coding region revealed expression of this receptor (data not shown).

3.1.1 HGF stimulates the Erk1/2 pathway and the Akt pathway in renal fibroblasts

For *in vitro* studies, FCS is used to activate HGF by proteolytic processing. However, FCS also contains growth factors, such as HGF and many others, which could interfere with the experiment [150, 151]. Therefore the minimal concentration of FCS, necessary to activate hHGF, was determined first. NRK49F cells were incubated in growth medium containing only 5 %, 2 %, 1 % and 0.5 % FCS, 24 h before stimulation. Commercially available hHGF was added in a concentration of 40 ng ml^{-1} for 15 minutes. As shown in figure 3.1.1, Erk 1/2 (p42/p44), a transducer of HGF signaling, was phosphorylated after hHGF treatment, while the non-stimulated cells cultured with 5 %, 2 % and 1 % FCS showed only faint phosphorylation. As the presence of 0.5 % FCS was sufficient to activate hHGF, subsequent analyses were performed in medium supplemented with 0.5 % FCS if not otherwise indicated.

Beside the FCS concentration also the concentration of hHGF for a sufficient stimulation of the cells had to be determined. Therefore NRK49F cells were incubated in medium containing 0.5 % FCS and were stimulated with hHGF concentrations ranging from 5 to 80 ng ml^{-1} for 15 minutes (Fig. 3.1.1 B). Phosphorylation of Erk1/2 in samples treated with 5 ng ml^{-1} was nearly as low as in controls but raised with increasing hHGF concentration and was most obvious at a concentration of 40 ng ml^{-1} (Fig. 3.1.1 B).

Taken together, 0.5 % FCS was sufficient to activate hHGF and a concentration of 40 ng ml^{-1} hHGF appeared to be most efficient for stimulation. Therefore, subsequent analyses were performed under these conditions.

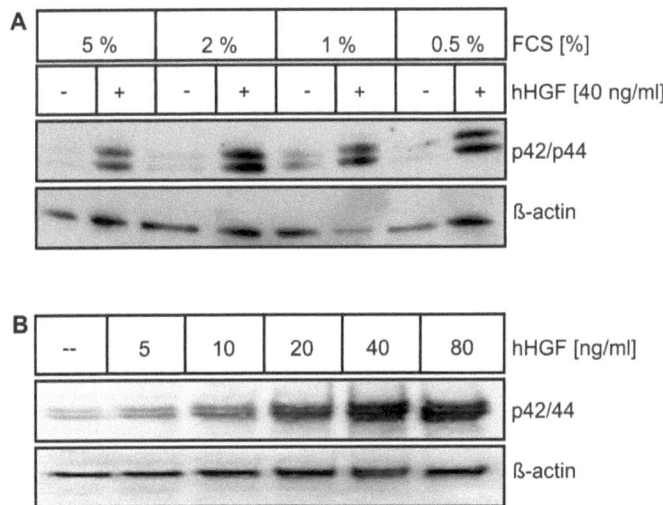

Fig. 3.1.1: Immunological detection of phosphorylated Erk1/2 (p42/p44) separated by SDS-gelelectrophoresis for the determination of (A) the necessary FCS concentration to activate hHGF and (B) the hHGF concentration to sufficiently stimulate NRK49F cells.
(A) NRK49F cells were incubated with DMEM supplemented with 5, 2, 1 and 0.5 % FCS. Cells stimulated with 40 ng ml^{-1} hHGF are indicated by (+), non-stimulated cells by (-). The protein bands for Erk1/2 (p42/44) with an apparent molecular weight of 42 and 44 kDa are shown. β-actin was used as loading control.
(B) NRK49F cells were stimulated with different hHGF concentrations ranging from 5-80 ng ml^{-1} hHGF for 15 minutes. Non-stimulated cells are shown as (--). The protein bands for Erk1/2 (p42/44) with an apparent molecular weight of 42 and 44 kDa are shown. β-actin was used as loading control.

Up to date, the anti-fibrotic effect of HGF is only reported to be mediated by counteracting the pro-fibrotic TGFβ [68, 69, 152]. But to clarify which signaling cascades are responsible for hHGF effects in renal interstitial fibroblasts, the phosphorylation statuses of Erk1/2 (p42/p44), Smad2, Stat3, and Akt (Fig. 3.1.2) after hHGF stimulation were analyzed. NRK49F cells were incubated in medium containing 0.5 % FCS. They were stimulated with 40 ng ml^{-1} hHGF for 5 and 15 min, as well as 1 and 6h.

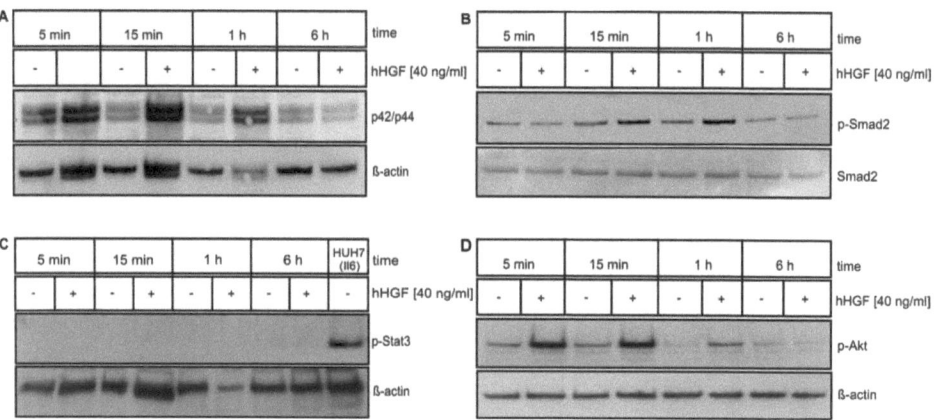

Fig 3.1.2 NRK49F cells were stimulated with hHGF [40 ng ml^{-1} for the indicated times. Cells stimulated with hHGF are marked by (+), non-stimulated cells by (-). The phosphorylation of (A) Erk1/2 (p42/44), (B) Smad2 (p-Smad2), (C) Stat3 (p-Stat3), and (D) Akt (p-Akt) were detected using phospho-specific antibodies. HUH7 cells stimulated with IL-6 served as positive control for Stat3-phosphorylation (C). β-actin (A, C, D) and Smad2 (B) were used as loading controls. The displayed analyses are representative for two independent experiments.

As illustrated in figure 3.1.2, hHGF treatment resulted in the phosphorylation of Erk1/2 (p42/44) already after 5 min, reached a peak between 15 and 60 min and then declined. Phosphorylation of Smad2 at the linker region was observable shortly later at 15 min after hHGF addition. The activation of Smad2 was still detectable 1 h after stimulation and returned to baseline 6 h after hHGF treatment (Fig. 3.1.2 B). Due to the fact that two other signaling cascades are reported to be turned on by HGF in epithelial cells, the Stat3 as well as the Akt pathway, the phosphorylation status of both was also checked. While Stat3 showed no phosphorylation by hHGF in renal fibroblasts (Fig. 3.1.2 C), the Akt protein was phoshorylated after hHGF treatment (Fig. 3.1.2 D). The onset of this pathway was simultaneous to the activation of the MAP/Erk cascade, demonstrating phosphorylation already 5 min after stimulation, which remained constant till 15 min, and then decreased.

3.1.2 Expression profiles induced by hHGF stimulation in renal fibroblasts

After identification of the signaling cascades that become turned on by HGF in renal fibroblasts, microarray analyses were performed to comprehensively identify the spectrum of genes that are affected by hHGF treatment. NRK49F cells were stimulated with a concentration of 40 ng ml^{-1} hHGF for 24 h. In contrast to the experiments concerning HGF signal transduction, examination of the mRNA levels of stimulated NRK49F cells was monitored in growth medium supplemented with 10 % FCS [153]. In the preceded experiments, the minimal amount of FCS was determined to detect only protein modification induced by exogenous hHGF. However, in this experiment the high amount of FCS was important to ensure proper metabolism and protein synthesis by the cells even over a long period of stimulation. Total RNA was isolated; the quality of the RNA preparations were verified by capillary electrophoresis on Bioanalyzer 2000 (Agilent) and subjected to an Affimetrix Chip for gene expression analyses. Un-stimulated NRK49F cells served as control. As shown in figure 3.1.3, gene expression profiling identified more than 1600 genes that were either up- or down-regulated by HGF. 58 up-regulated genes displayed a fold change higher than 3, however, most of the more than 3-fold differentially expressed genes were down-regulated (n = 202) (Fig. 3.1.3 A).

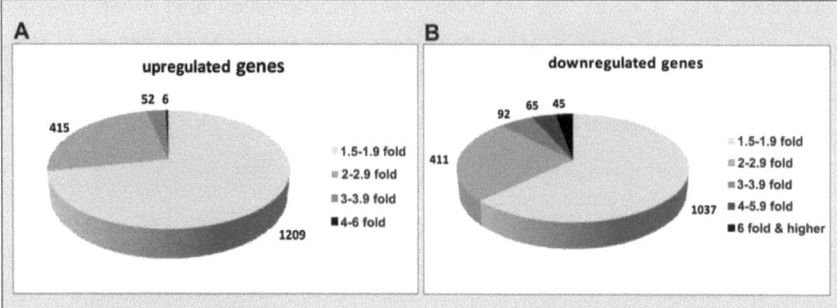

Fig. 3.1.3 Expression profiling by microarray analyses of hHGF stimulated NRK49F cells. Up-regulated (A) and down-regulated genes (B) after treatment with hHGF for 24 h. Total RNA was extracted and microarray analyses were performed.

Functional clustering of the genes revealed the up-regulated genes to be preferentially involved in motor and intracellular transporter activity (Fig. 3.1.4 A). The down-regulated genes, however, were linked to ECM production and degradation, cell proliferation, immune response and signal transduction (Fig. 3.1.4 B and supplemental table S1).

RESULTS

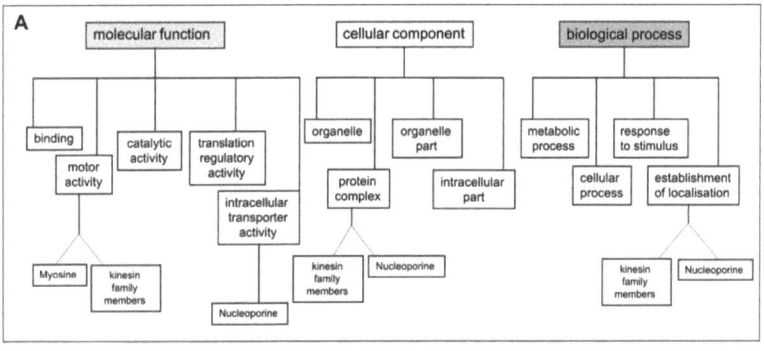

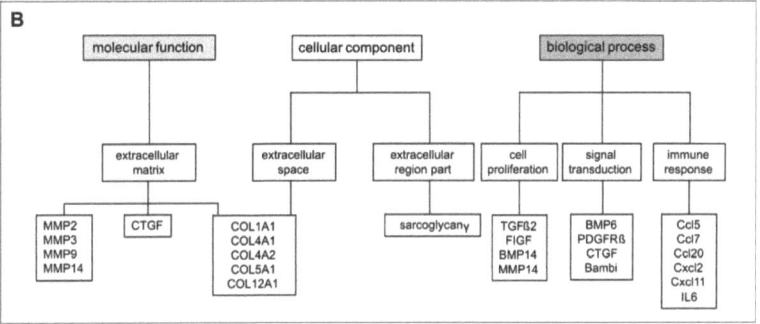

Fig. 3.1.4: Genes identified by microarray analyses were classified in three groups (biological process, molecular function and cellular component) according to the GO-terms. (A) genes that were up-regulated by hHGF and (B) genes that were down-regulated.

In order to identify genes affected by hHGF that are involved in the mechanisms of fibrogenesis, here the main interest was focussed on candidates selected by their pronounced divergent expression profile and their potential association with pro-fibrotic processes. 16 genes were chosen for further validation by real-time PCR and the selected genes were subdivided into 4 groups:

(1) signal transducers linked to specific fibrotic signaling cascades
(2) intracellular mediators of TGFβ signals, the Smad genes
(3) members of the CCN family
(4) fibrotic markers and collagens

3.1.2.1 HGF inhibits expression of signal transducers linked to fibrotic processes

To verify the results obtained by gene chip analyses, real-time PCR was performed on expression levels of selected genes. The first analyzed group involved genes that are linked to signaling cascades. The fibrotic mediator TGFβ was more than 40 % down-regulated by hHGF validated by real-time PCR (Fig. 3.1.5 A). Likewise the BMP and activin membrane-bound inhibitor (Bambi) displayed a 53 % decreased transcript level (Fig. 3.1.5 A). In addition, the expression level of the receptor PDGFRβ (platelet-derived growth factor receptor beta) was 70 % reduced. PDGF signaling via this receptor has been implicated in several fibrotic conditions and is assumed to play a role in driving proliferation of cells with a myofibroblastic phenotype [154]. A dramatic decrease by hHGF was also observable for FIGF (c-fos induced growth factor), a gene related to the platelet-derived growth factor/vascular endothelial growth factor family [155]. This factor is also known as VEGF-D and the presence of hHGF strongly diminished transcript levels up to 92 % (Fig. 3.1.5 A).

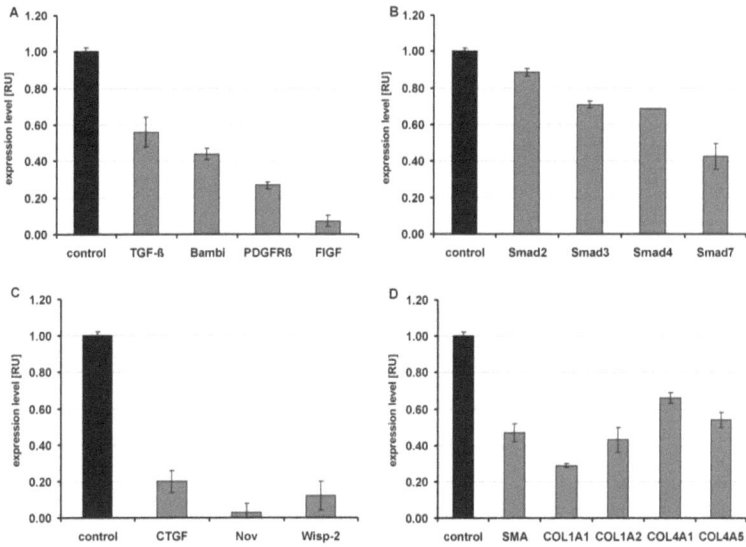

Fig. 3.1.5: Expression analyses of NRK49F cells treated with hHGF for 24 h. Real-time PCR was performed with primers specific for (A) TGFβ, Bambi, PDGFRβ and FIGF, (B) Smad2, Smad3, Smad4 and Smad7, (C) CTGF, Nov, Wisp-2 and (D) for SMA, COL1A1, COL1A2, COL4A1 and COL4A5. NRK49F cells were cultured in the presence of hHGF [40 ng ml^{-1}] for 24 h (grey bars). The expression levels were normalized to HPRT and compared to the untreated cells that served as control (black bars). Each bar represents the mean and SD of three independent experiments. The control value was arbitrarily set 1.

3.1.2.2 Effects of HGF on the expression level of Smad genes

Smad-signaling is one of the crucial pathways involved in renal fibrosis [156]. Therefore the effect of hHGF on the expression levels of different Smads was also validated by real-time PCR. As demonstrated in figure 3.1.5 B, HGF treatment had only slight effects on the mRNA level of Smad2 (10 % reduction) or Smad3 and Smad4 (30 % reduction). The inhibitor Smad7, however, known to be induced by a feed-back mechanism of Smad2/3 signaling [33], displayed a highly reduced mRNA level of almost 60 % compared to the control level (Fig. 3.5.1 B).

3.1.2.3 Effects of HGF on the expression level of CCN family members

A strong down-regulation by hHGF was observable for members of the CCN family, CTGF (connective tissue growth factor), Wisp-2 (wnt-induced secreted protein-2) and Nov (nephroblastoma overexpressed protein). All of them are extracellular matrix-associated proteins that amongst others play critical roles in injury repair, fibrotic diseases and cancer [157]. Stimulation of NRK49F cells with hHGF for 24 h resulted in a significant decrease of the expression levels of all three CCN genes with the strongest effect on Nov expression (Fig. 3.1.5 C). While the transcript levels of CTGF were reduced up to 80 %, the diminished Nov expression (96 %) exceeded the 90 % reduced Wisp-2 expression.

3.1.2.4 Effects of HGF on the expression level of fibrotic markers and collagens

The actin isoform of smooth muscle actin (SMA) is a marker protein for myofibroblastic differentiation and prominently linked to fibrotic processes [10]. Expression analyses by real-time PCR demonstrated that the transcript level of SMA was narrowed to 55 % by hHGF treatment. Furthermore, subunits of collagen I and IV were also negatively affected by hHGF. The main matrix protein accumulated during fibrosis is collagen I. Real-time PCR revealed a down-regulation of the mRNA levels of the collagen Iα1 subunit (70 % reduction) as well as the collagen Iα2 subunit (57 % reduction). Likewise the mRNA levels of the α1 (34% reduction) and α5 subunit (46 % reduction) of collagen IV were highly reduced by hHGF.

Taken together these results demonstrate that HGF resulted in a down-regulation of genes that are particularly involved in the synthesis of extracellular matrix (collagenes), extracellular matrix-associated genes that amongst others play critical roles in injury repair, fibrotic diseases and cancer (CCN genes) as well as genes that are linked to signaling cascades that are implicated in several fibrotic conditions. Thus, the results of the microarray analyses could be validated and moreover a more accurate expression profile of selcted candidate genes was determined.

But with regard to the analyses of the signaling cascades that are turned on by hHGF in interstitial fibroblasts (3.1.1), it remains still unclear if hHGF regulates the selected genes via the onset of the MAP/Erk pathway or the activation of the Akt pathway. Therefore, siRNA approaches to block either the Smad pathway or the Akt pathway were performed.

3.1.2.5 Smad independent anti-fibrotic effects of HGF

In order to investigate whether genes were repressed in response to hHGF by inhibited Smad signaling due to linker phoshorylation after Erk1/2 activation, or by activation of the Akt pathway, NRK49F cells were transfected with siRNA specific for Smad4 and specific for Akt, respectively. In both cases, cells were also additionally stimulated with hHGF [40 ng ml^{-1}] for 24 h. Then, the 16 previously identified hHGF target genes were analyzed by quantitative real-time PCR. For control, scrambled siRNA was transfected under the same conditions.

The siRNA transfection for Smad4 reduced the mRNA level of Smad4 to more than 80 % while the silencing of Akt resulted in more than 60 % reduced transcript levels (data not shown). All of the selected genes validated by real-time PCR (3.1.2.1-3.1.2.4) were also analyzed in regard to their mRNA expression levels after knockdown of Smad4 or Akt. However, only genes that showed an additional effect by hHGF treatment after down-regulation via Smad4 silencing are displayed in figure 3.1.6. Correspondent to Smad4 inhibition these genes are also displayed for the Akt silencing experiments (Fig. 3.1.7).

With regard to the blockade of the Smad pathway by silencing Smad4 via specific siRNA, the expression levels of TGFβ, Smad2 and Smad3 were not affected by the knockdown of Smad4 (data not shown). In contrast, Smad7, Bambi, SMA, COL1A1, COL1A2, COL4A5, CTGF, Nov, Wisp-2, PDGFRβ, and FIGF displayed highly reduced mRNA expression levels after Smad4-blockade by siRNA (shown for Nov, Wisp-2, CTGF, PDGFRβ, FIGF, and COL1A1 in Fig. 3.1.6). Furthermore, HGF stimulation of Smad4 silenced cells resulted in an additional decrease in the mRNA levels of Nov, Wisp-2, PDGFRβ, FIGF, and CTGF (Fig. 3.1.6 A-E). The other genes were not further affected by hHGF, as demonstrated for COL1A1 (Fig. 3.1.6 F).

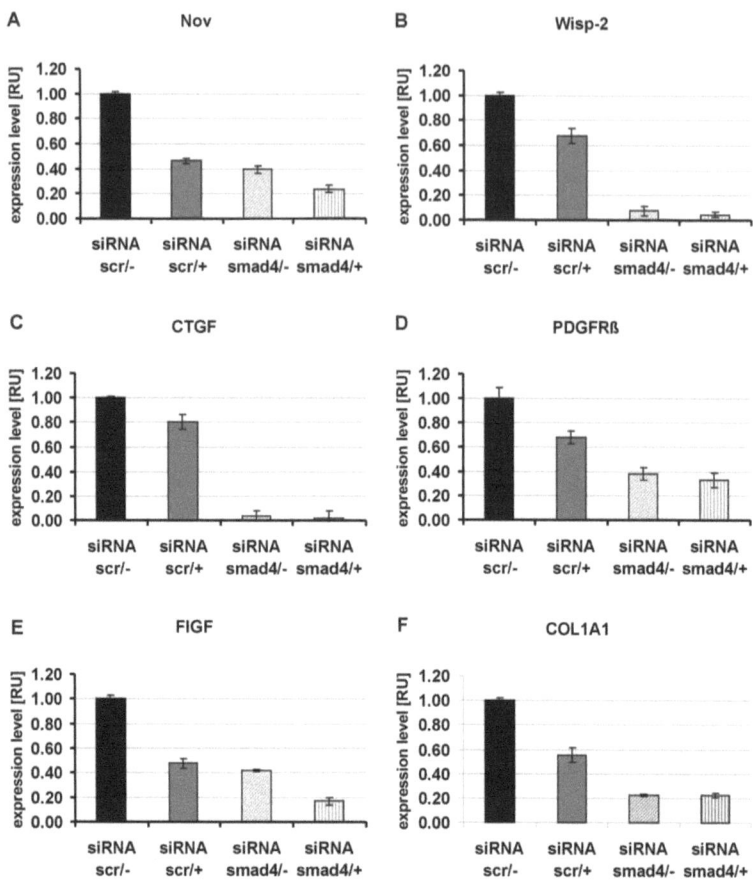

Figure 3.1.6: Effects of hHGF on the expression levels of different genes in Smad4-inhibited NRK49F cells. Expression levels of (A) Nov, (B) Wisp-2, (C) CTGF, (D) PDGFRβ, (E) FIGF, and (F) COL1A1 are displayed. Cells were transfected with either Smad4 siRNA (hatched bars) or scrambled (scr) control siRNA (black and grey bars), respectively. Twelve hours after transfection, medium was exchanged with medium containing 10 % FCS and cultivated for additional 12 h. Then the cells were treated with (+) or without hHGF (-) [40 ng ml^{-1}] for another 24 h. RNA was extracted, reverse transcribed and mRNA expression was determined by real-time PCR. All expression levels were normalized to HPRT. The graphics are representative for three independent experiments. Untreated cells served as control (black bar) and the value of the control was arbitrarily set 1.

The additive effect of hHGF on Smad4 silenced cells was directly compared to un-stimulated Smad4 inhibited cells as shown and summarized in figure 3.1.7. FIGF displayed with 60 % down-regulation of the mRNA level the highest reduction by hHGF in comparison to Smad4 silenced cells. But also the further reduced expression level of Nov (40 %), Wisp-2 (41 %), CTGF (32 %), and PDGFRβ (14 %) suggest an additional Smad independent contribution by hHGF.

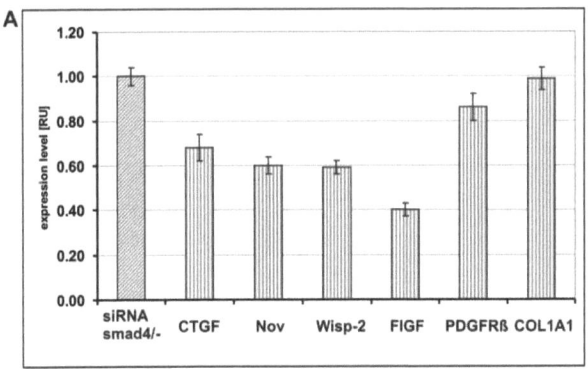

Figure 3.1.7: Additional down-regulation by hHGF in Smad4-silenced NRK49F cells compared to un-stimulated Smad4 silenced cells. (A) The expression levels of Nov, Wisp-2, CTGF, PDGFRβ, FIGF, and COL1A1 are displayed. Cells were transfected with Smad4 siRNA. Twelve hours after transfection, medium was exchanged with medium containing 10 % FCS and cultivated for additional 12 h. Then the cells were treated with 40 ng ml^{-1} hHGF for another 24 h (hatched bars). RNA was extracted, reverse transcribed and mRNA expression was determined by real-time PCR. All expression levels were normalized to HPRT. The graphic is representative for three independent experiments. Un-stimulated Smad4 silenced cells served as control (black bar) and the value of the control was arbitrarily set 1. The additional reduction of the transcript levels after hHGF treatment is indicated in percentage. P-values < 0.05 were considered significant (B).

To clarify whether the additional repression of the mRNA levels could be attributed to a regulation via the Akt pathway, real-time PCR of the transcript levels in cells that were silenced for Akt were performed (fig. 3.1.8).

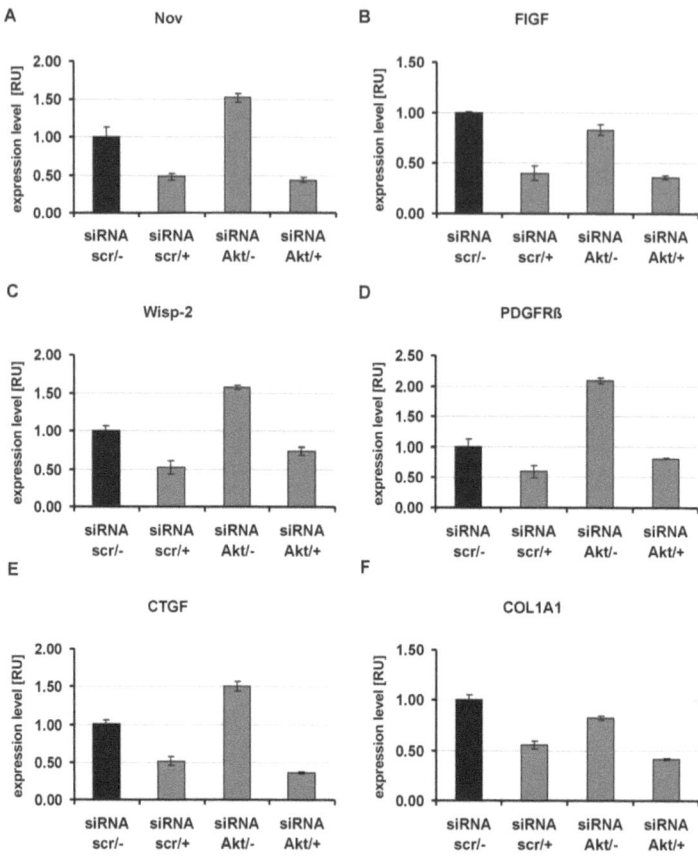

Figure 3.1.8: Effects of hHGF on the expression levels of different genes in Akt-inhibited NRK49F cells.
The expression levels of Nov (A), Wisp-2 (B), CTGF (C), PDGFRβ (D), FIGF (E) and COL1A1 are displayed. Cells were transfected with either Akt siRNA (Akt) or scrambled (scr) control siRNA (black and grey bars), respectively. Twelve hours after transfection, medium was changed to medium containing 10 % FCS and cultivated for additional 12 h. Then the cells were treated with (+) or without (-) 40 ng ml^{-1} hHGF for another 24 h. RNA was extracted, reverse transcribed and mRNA expression was determined by real-time PCR. All expression levels were normalized to HPRT. The graphics are representative for three independent experiments. Untreated cells served as control (black bar) and the value of the control was arbitrarily set 1.

Repression of Akt signaling resulted in a dramatic increase of Nov, Wisp-2, CTGF, and PDGFRβ mRNA levels, indicating that Akt plays a decisive role in the regulation of these genes. Furthermore, additional treatment with hHGF demonstrated a high down-regulation of the expression levels, which is most likely due to the inhibition of the Smad pathway. However, FIGF and COL1A1, showed no altered expression after Akt knockdown, but a strong repression by the additional stimulation with hHGF. The other genes of the 16 selected candidates showed no alteration in their expression levels after Akt knockdown.

In summary, these data provide evidence that HGF stimulation of renal fibroblasts results in activation of both, the Erk1/2 and the Akt pathway. Functional cluster analyses and quantitative real-time PCR assays indicate that the HGF-stimulated pathways transfer the anti-fibrotic effects in renal interstitial fibroblasts by reducing expression of extracellular matrix proteins, various chemokines, and members of the CCN family. Blocking of the Smad pathway by RNA interference revealed that not only the interaction with the Smad pathway by HGF is involved in the down-regulation of fibrotic mediators but also the HGF stimulated Akt pathway.

Thus, HGF has a broad effect on the inhibition of fibrotic markers and mediators in renal interstitial fibroblasts. Based on these *in vitro* results, further analyses were addressed to the anti-fibrotic role of hHGF in interstitial renal fibrosis *in vivo*. Therefore a mouse model for interstitial fibrosis was used and hHGF was administered using the adeno-associated virus as gene vehicle.

3.2 The anti-fibrotic function of HGF in a gene therapeutical approach, *in vivo*

To examine the anti-fibrotic effect of hHGF also *in vivo*, the COL4A3 knockout mouse model was used. COL4A3 knockout mice are a model for the human Alport syndrome, representing the autosomal form. These mice are homozygous for the deletion of the α3 subunit of collagen IV [16]. The disease progression is very similar to that reported in studies of humans and include hearing defects, microhematuria, proteinuria, and irregular thickening and splitting of the glomerular basement membrane [16]. Mice suffering from the Alport syndrome finally develop an interstitial fibrosis and based on the genetic background (129/Sv) these mice develop endstage renal failure within 14 weeks [16]. The morphology of the kidney is characterized by thickening and splitting of the glomerular basement membrane. The thickening starts with the age of 4 weeks in the external capillary loops and spreads out in the whole kidney. The mice die by renal failure at the age of 11-14 weeks.

For the application of hHGF the adeno-associated virus (AAV) was chosen as gene vehicle, caused by the fact that this vector is non-pathogenic and non-immunogenic and is able to transduce both, proliferating and quiescent cells [91, 92]. Furthermore, AAV allows a cell- and tissue specific application as well as a long-term expression of the transgene by site-specific integration into the genome or episomal persistence [93].

3.2.1 Renal transduction efficiency of different AAV serotypes

The adeno-associated virus (AAV) has become a versatile vector platform due to the availability of a wide spectrum of serotypes, mosaic and hybrid vectors as well as tailored mutants. In order to investigate the potential of this vector system in the treatment of renal interstitial fibrosis, the commonly used serotype rAAV2 was compared with two less characterized serotypes, namely rAAV-8 and rAAV-9, in their capability to efficiently transduce renal epithelial cells.

3.2.1.1 Low *in vitro* transduction of renal epithelial cells by rAAV2

To investigate whether rAAV2 is a suitable vector for gene transfer into kidney epithelial cells *in vitro*, a reporter-construct containing the Enhanced-Green-Fluorescence-Protein (EGFP) under the control of the ubiquitously active cytomegalovirus (CMV) promoter was packaged into rAAV2-capsids (rAAV2-GFP). Rat kidney epithelial cells (NRK52E) and rat kidney fibroblasts (NRK49F) were exposed to 1×10^4 particles/cell for 48 h and reporter gene expression was examined under a fluorescence microscope (Fig.3.2.1). rAAV2-permissive HeLa cells served as positive control.

RESULTS

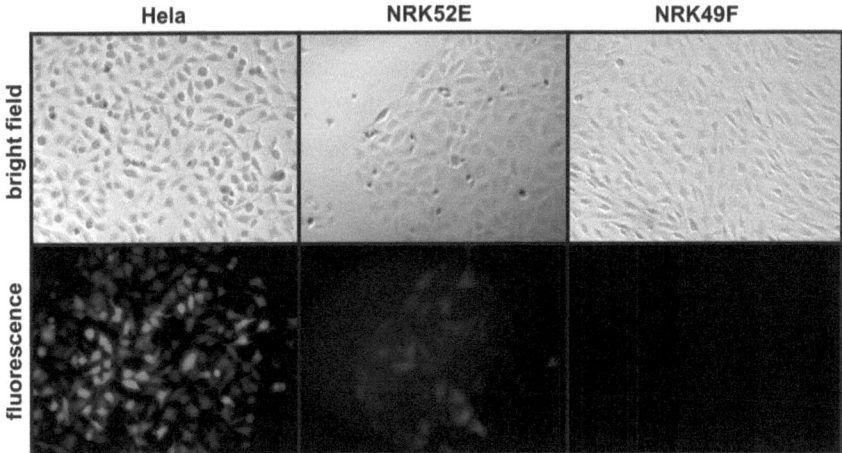

Fig. 3.2.1: *In vitro* transduction of HeLa, NRK52E and NRK49F cells by rAAV2. Cells were transduced with 1×10^4 particles/cell for 48 h and GFP-expression was examined under a fluorescence microscope. The results are representative for two independent experiments.

While rAAV2-permissive HeLa cells exhibited high GFP expression, transduction of NRK52E cells showed a low expression of GFP, indicating low efficiency of rAAV2-mediated gene transfer into renal epithelial cells. NRK49F cells, however, showed no GFP-fluorescence at all. These findings led to the assumption, that the rAAV2 is able to transduce renal epithelial cells but with relatively low efficiency. Renal fibroblasts, however, are not transduced by rAAV2 (fig 3.2.1).

Until now there is no available cell line to determine the transduction efficiency of rAAV8 and rAAV9 *in vitro*. Therefore the experiments with these serotypes were performed exclusively *in vivo*.

3.2.1.2   *In vivo* transduction of liver and kidney by rAAV2, rAAV8 and rAAV9

In order to identify the best suited AAV serotype that mediates renal expression of HGF *in vivo* following intravenous administration, the three serotypes, rAAV2, rAAV8 and rAAV9, were analyzed with regard to their ability to target the kidney. All three serotypes contained the GFP reporter controlled by the ubiquitously active cytomegalovirus (CMV) promoter. The expression cassettes were flanked by the serotype 2 packaging signals (ITR), thus, rAAV serotype 8 and 9 were packaged as pseudo-types. Equivalent numbers of vector genomes (5×10^{11} vector genomes) were administered by a single intravenous infusion via the tail vein of 6-week old male COL4A3-knockout mice (n = 6 animals per cohort). The mice were sacrificed 2 weeks after gene delivery and liver and kidneys were taken for cryo-sections as well as paraffin embedding.

To determine the amount of virus particles which have successfully transduced liver and kidney cells, real-time PCR of DNA was performed. In addition, real-time PCR of reverse transcribed RNA was used to define the GFP expression rate in both organs. Figure 3.2.2 represents the quantitative analysis of the vector biodistribution (A, B) and the expression level (C, D) in both organs following intravenous administration of rAAV2-GFP, rAAV8-GFP and rAAV9-GFP.

RESULTS

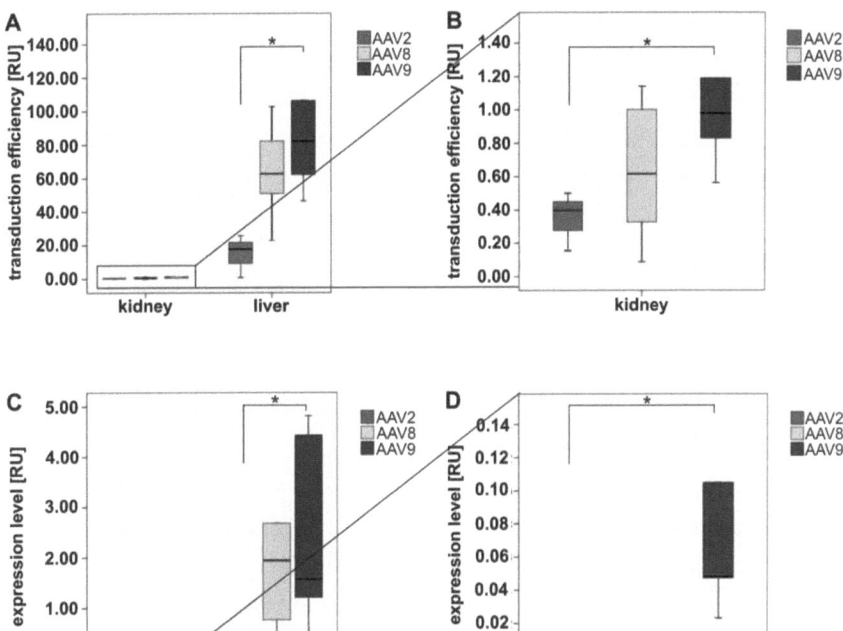

Fig. 3.2.2: Transduction efficiency (A, B) and GFP expression level (C, D) of rAAV2, rAAV8 and rAAV9 in liver and kidney analyzed by real-time PCR after systemic administration of rAAV2-GFP, rAAV8-GFP and rAAV9-GFP via the tail vein. 5×10^{11} particles were injected into male COL4A3 knockout mice (6 mice per group). Mice were sacrificed two weeks post injection and DNA (for transduction efficiency) and RNA (for expression analyses) were extracted. The transduction efficiency (RU = relative units) was normalized to mCRP (A, B). RNA was transcribed into cDNA and the GFP expression rate (RU = relative units) was determined and normalized to HPRT (C, D). $P < 0.05$ were considered statistically significant, indicated by *. B and D are rescaled figures derived from A and C, respectively.

With regard to the transduction efficiency, all three serotypes showed a preference for liver tissue with rAAV8 and rAAV9 being clearly superior in transduction (Fig. 3.2.2 A). rAAV2-GFP showed the lowest transduction of the liver with around 20 relative units (RU). rAAV8-GFP exceeded this transduction rate three times with 60 RU. The highest transduction of the liver could be observed with 80 RU in rAAV9 injected mice.

Although targeting the kidney by all three serotypes was less efficient (50-100 times) than targeting the liver, a diagram using a smaller scale demonstrated predominant renal transduction by rAAV8 and rAAV9 (Fig. 3.2.2 B) with the most efficient transduction of the kidney by rAAV9 (Fig. 3.2.2 B). For rAAV2, only low numbers of vector genomes could be detected within the kidney.

As rAAV9 showed the highest transduction of the liver and the kidney the next intention was to determine the expression of GFP in the liver and the kidney. Therefore, both, real-time PCR on cDNA (Fig. 3.2.2 C and D) and immunohistochemical stainings for GFP on paraffin sections (Fig. 3.2.3), were performed.

As expected from the above mentioned results, quantitative real-time PCR showed considerably higher GFP reporter expression in the livers and the kidneys by rAAV8 and rAAV9 (Fig. 3.2.2 C). According to the DNA analyses, demonstrating a predominant transduction of the livers, noticeable higher GFP expression was detectable in the livers (except rAAV2-GFP) compared to the kidneys (Fig. 3.2.2 C). To allow a closer look at the GFP expression level of the different serotypes in the kidneys, figure 3.2.2 D displays the reporter gene expression level of the kidneys plotted in a smaller scale. rAAV9 injection resulted in the highest reporter gene expression in renal cells, while rAAV2 as well as rAAV8 achieved only low GFP expression. Also the liver showed the highest GFP expression by rAAV9.

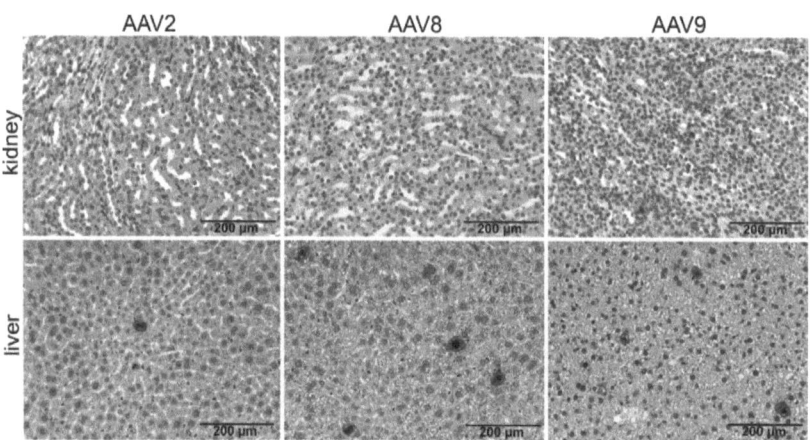

Fig. 3.2.3: GFP expression in kidney and liver by rAAV2-GFP (AAV2), rAAV8-GFP (AAV8) and rAAV9-GFP (AAV9) were visualized by immunohistochemical GFP staining. Equivalent vector genomes (5×10^{11}) of each AAV vector were administered by a single intravenous infusion via the tail vein in male COL4A3 knockout mice (6 mice/group). Mice were sacrificed 2 weeks post injection and paraffin-sections of kidneys and livers were immunohistochemically stained for GFP and counterstained with hematoxyline. Sclae bars are indicated.

These differences in the expression levels were also obvious in the histological immunostainings for GFP (Fig. 3.2.3).

In agreement to the quantitative real-time PCR results, GFP immunostaining after rAAV2 transduction showed only few liver cells GFP-positive whereas no single kidney cell expressing GFP could be detected. Histological examination of the livers and kidneys of rAAV8-GFP and rAAV9-GFP transduced mice revealed a similar or even slightly higher GFP expression by rAAV9 in the liver and a defined GFP-expression of the kidney by rAAV9. While the immunohistochemistry of rAAV2 and rAAV8 transduced mice showed no renal GFP-expressing cell, the immunohistochemistry of rAAV9 injected mice displayed several GFP-positive epithelial cells of the proximal and distal tubuli, predominantly scattered over the medulla, which is the inner part of the kidney (Fig. 3.2.3).

Thus, the data represent a high transduction efficiency of the liver by the three serotypes, rising from rAAV2 to rAAV9 and a low efficiency of the kidneys by all three serotypes. But again, rAAV9 showed the highest transduction rate for renal gene delivery. Moreover, GFP-expression in the kidney was predominantly observed in the innermost part of the kidney, the medulla.

As the reporter gene analyses demonstrated a preference for liver tissue with rAAV8 and rAAV9 being clearly superior in transduction (Fig. 3.2.2 A) and transgene expression, in this study, the use of a tissue-specific promoter should restrict the expression to the kidney by driving specific expression in renal tubular epithelial cells. Therefore, a mammalian promoter was used for the expression of HGF that was reported to be kidney-specific *in vitro* and *in vivo* [158-160].

3.2.2 Transgene expression limited to renal tubuloepithelial cells

To restrict transgene expression to the kidney, two constructs with the kidney-specific promoter Ksp-cadherin, including different enhancer elements, were generated. Ksp-cadherin is a cell-adhesion molecule that is reported to be exclusively expressed in tubular epithelial cells in the kidney and in the developing genitourinary tract [158]. Shao and colleagues have shown that a 1342-bp fragment of the 5'-region of the Ksp-cadherin gene contains all necessary elements driving gene expression in renal tubular epithelial cells [158]. Therefore, this Ksp-promoter was used in this study and cloned with two different enhancers (CMV-enhancer or beta-globin minimal promoter (beta-globin MM)) in front of a firefly-luciferase reporter for functional analyses of the promoter activity (Fig. 3.2.4 A). For construction details see supplemental figure S1 (pGL-Ksp-MM-Luc) and supplemental figure S2 (pGL-CMV-Ksp-Luc) in the attachment.

3.2.2.1 Transgene expression driven by the Ksp-promoter in renal tubuloepithelial cells *in vitro*

For functional analyses, the two generated luciferase reporter plasmids, pGL-CMV-Ksp-Luc (Fig. 3.2.4 A (2)) and pGL-Ksp-MM-Luc (Fig. 3.2.4 A (3)) were transfected into renal epithelial cells (NRK52E) using cationic liposomes (Lipofectamin, Invitrogen, GER). A third plasmid containing the firefly-luciferase regulated by the ubiquitously active CMV-promoter (CMV-Luc) (Fig. 3.2.4 A (1)) was transfected as positive control. Co-transfection in all three cases with 1/10 of the pRL-TK vector (Promega) encoding the Renilla-luciferase were performed as control for transfection efficiency. 48 h post transfection the cells were lysed and assayed for luciferase-activity.

Since promoter activity should be kidney-specific, the reporter plasmids as well as the CMV-Luc plasmid were transfected into human hepatoma cells (HUH7) as control for cell-specificity. Figure 3.2.4 B represents the relative luciferase-activity in percentage measured in NRK52E and HUH7 cells following transfection.

RESULTS

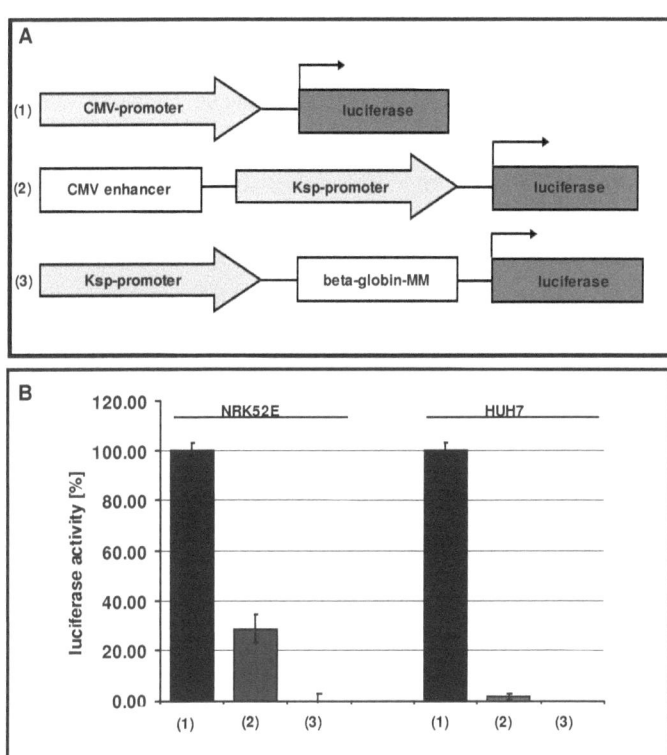

Fig. 3.2.4 (A) Luciferase reporter constructs. (1) CMV-Luc: luciferase regulated by the CMV promoter, (2) pGL-CMV-Ksp-Luc: luciferase under the control of the Ksp-promoter enhanced by the CMV enhancer and (3) pGL-Ksp-MM-Luc: luciferase regulated by the Ksp-promoter enhanced by the beta-globin minimal promoter.
(B) Luciferase activity of different promoter constructs in cultured renal epithelial cells (NRK52E) and hepatocytes (HUH7). Cells were transfected with pGL-CMV-Ksp-Luc (pink bar), pGL-Ksp-MM-Luc or CMV-Luc (black bars) as positive control. Luciferase activity was measured after 48 h. To control transfection efficiency, cells were co-transfected with 1/10 of the vector pRL-TK vector and luciferase activity was normalized to renilla-luciferase activity. The luciferase activity of the positive control (CMV-Luc) was set on 100 %. Data are representatives for three independent experiments.

The positive control, expressing luciferase via the ubiquitously active CMV-promoter, displayed the highest relative luciferase activity. This activity was set 100 % (black bar). Transfection of CMV-Ksp-Luc into NRK52E cells reached one third of the relative luciferase activity compared to the highly active positive control. The Ksp-

MM-Luc transfected cells showed no luciferase activity at all, indicating no activity of the Ksp-promoter enhanced by the β-globin minimal promoter.

For the hepatoma cells (HUH7) the Ksp-MM-Luc construct showed again no luciferase activity, while the CMV-Ksp-Luc reporter construct resulted in a marginal luciferase activity (Fig. 3.2.4 B).

These results demonstrate that *in vitro* the Ksp-cadherin promotor driven by the CMV enhancer resulted in a high expression rate in kidney epithelial cells whereas the Ksp-cadherin promoter enhanced by the β-globin minimal promoter seems to be inactive *in vitro*.

3.2.2.2   Reporter gene expression driven by the kidney-specific Ksp-promoter *in vivo*

To analyze the promoter construct of the kidney-specific promoter and the CMV enhancer *in vivo*, reporter gene analyses using GFP were performed. Therefore, an expression cassette containing GFP under the control of the Ksp-promoter enforced by the CMV enhancer (CMV-Ksp-GFP; see supplemental figure S3) was packaged into rAAV9 capsids and six week old male COL4A3 knockout mice (6 mice) were injected with 5×10^{11} particles of either rAAV9-CMV-Ksp-GFP or rAAV9-CMV-GFP (see 3.2.1.2) and sacrificed 2 weeks post injection. Organs were taken for DNA, RNA as well as immunohistological analyses.

Although the same numbers of viral particles were injected, the transduction efficiency of the CMV-Ksp-GFP vector was much better (data not shown).

The immunohistological stainings for GFP showed strong GFP signals in the kidney (Fig. 3.2.5). But also the liver showed a high expression of GFP. Obviously, in contrast to previous reports that demonstrated the Ksp-promoter to be exclusively expressed in tubular epithelial cells in the kidney [158] the Ksp-promoter, enforced by the CMV enhancer, did not restrict expression to the kidney. But, to verify if this promoter construct would enable a controllable expression for combined paracrine and endocrine HGF delivery, further organs were analyzed with regard to GFP expression. Interestingly, the immunohistochemical stainings for GFP expression could show a tissue restriction in transgene expression by the use of the Ksp-promoter construct. Neither in spleen and lung, nor in heart GFP expression was detectable (Fig. 3.2.5).

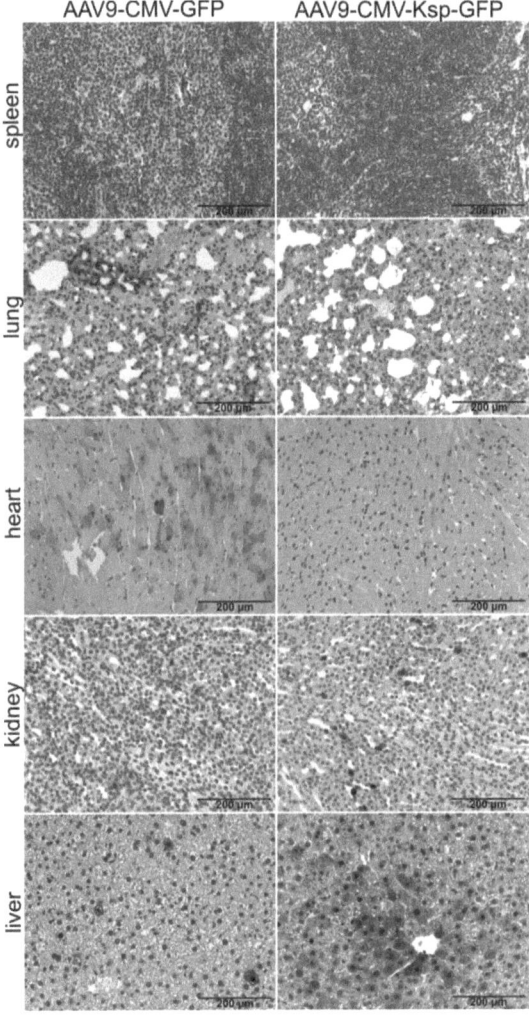

Fig. 3.2.5: GFP expression in spleen, lung, heart, kidney and liver by rAAV9-CMV-GFP and rAAV9-CMV-Ksp-GFP, visualized by immunhistochemical GFP staining. Equivalent vector genomes (5×10^{11}) of each AAV vector were administered by a single intravenous infusion via the tail vein in 6 week old male COL4A3 knockout mice (6 mice/group), however, transduction efficiency was less in rAAV9-CMV-GFP (data not shown). Mice were sacrificed 2 weeks post injection and paraffin-sections of spleen, lung, heart, kidney and liver were immunohistochemically stained for GFP and counterstained with hematoxyline. Scale bars are indicated.

In summary, the generated kidney-specific reporter-construct indeed resulted in a restricted expression of the transgene, however, a complete restriction to the kidney was not achieved. Instead, this promoter construct mediated expression in both organs, the kidney and the liver, with a greater extend in the liver. Therefore, the gene therapeutical approach using hHGF as transgene was performed by targeting the liver parenchyma and the renal tubuloepithelium for combined endocrine and paracrine transgene delivery.

3.2.3 AAV-induced recombinant hHGF expression as a gene therapeutical approach for the treatment of tubulo-interstitial fibrosis *in vivo*

As reporter gene analyses of the chosen promoter construct still revealed an expression in the liver and the kidney, the gene therapeutical approach for renal interstitial fibrosis using HGF as transgene was performed as a bidirectional approach. Although the intention was to restrict transgene expression to the kidney, a possible advantage of such an approach could be a local HGF expression at the place of damage that is supported by an additional HGF supply via the blood circulation.

Therefore, the human hepatocyte growth factor (hHGF) was now utilized as transgene in a gene therapeutical approach aimed to arrest or improve renal interstitial fibrosis.

3.2.3.1 Construction of the hHGF-expression cassette for the generation of AAV8 and 9 vectors

To examine the anti-fibrotic effect of HGF in the kidney *in vivo*, a hHGF-expression cassette that restricts expression to the liver and the kidney was generated. Based on the functional analyses of the two different promoter constructs in 3.2.2 and the functional analyses of the KSp promoter enhanced by the CMV enhancer *in vivo* (3.2.2.2), the expression of hHGF was placed under the control of the Ksp-cadherin promoter, strengthened by the CMV enhancer. The already generated Ksp-cadherin promoter construct including the luciferase-gene as reporter (CMV-Ksp-Luc; see

3.2.2), was used as backbone. The luciferase-gene with the poly(A) was excised and replaced by the hHGF coding region with a poly(A) stretch resulting in the plasmid pGL-CMV-Ksp-hHGF (see supplemental figure S4).

For the construction of the kidney-specific hHGF-expression cassette for AAV packaging, the expression cassette was cloned into the pSUB-201 [161]. This vector contains all AAV-2 wild-type coding regions and the *cis* acting terminal repeats that are required for recombinant virus production. The generated vector was named pSUB201-CMV-Ksp-hHGF (Fig. 3.2.6 and supplemental Fig. S5).

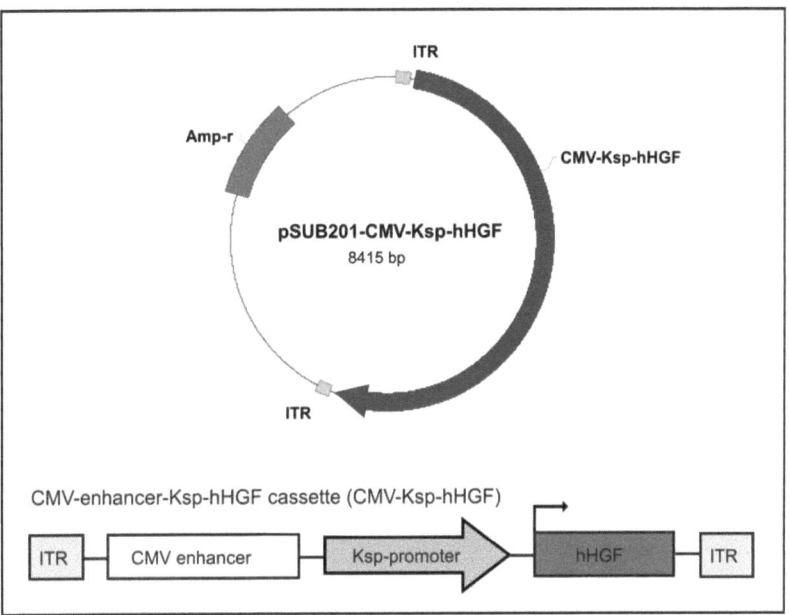

Fig. 3.2.6 hHGF-construct packaged into rAAV-8 and rAAV-9 for injection of COL4A3 knockout mice. hHGF expression is driven by the kidney-specific promoter Ksp-Cadherin (Ksp-promoter), enhanced by the CMV enhancer. The cassette was cloned into the pSUB-201 backbone, that contains the inverted terminal repeats (ITRs), the only *cis* acting sequences required for recombinant virus production.

3.2.3.2   *In vivo* administration of hHGF by rAAV8 and rAAV9

To evaluate the anti-fibrotic mechanisms of hHGF and the pattern of gene expression after hHGF administration in COL4A3 knockout mice developing interstitial kidney fibrosis, the kidney-specific hHGF-rAAV vectors serotype 8 and 9 were systemically administered into male COL4A3 knockout mice via the tail vein. 4 week old mice were transduced with 5 x 10^{11} particles rAAV8-CMV-Ksp-hHGF (n = 8) or rAAV9-CMV-Ksp-hHGF (n = 8) and sacrificed 5.5 weeks after. Blood was taken for serum analyses, while kidneys and livers were dissected for cryo-sections and paraffin embedding. Mice injected with empty capsids (n = 6) served as controls.

For the analyses of the hHGF sera levels of the rAAV8-CMV-Ksp-hHGF and rAAV9-CMV-Ksp-hHGF transduced mice a hHGF specific ELISA was performed. As demonstrated in figure 3.2.7 there was no hHGF detectable in the sera from mice that received empty capsids as control. In contrast, serum samples of rAAV8-CMV-Ksp-hHGF transduced mice displayed around 100 pg ml^{-1} hHGF expression that was exceeded by the rAAV9 transduced mice with a mean value of 340 pg ml^{-1} hHGF (Fig. 3.2.7). Even though, there was a variance in the hHGF sera level of the rAAV9-CMV-Ksp-hHGF transduced mice, the sera levels were significantly higher compared to the rAAV8-CMV-Ksp-hHGF injected mice. This variance is presumably due to an inappropriate application of the vector genomes.

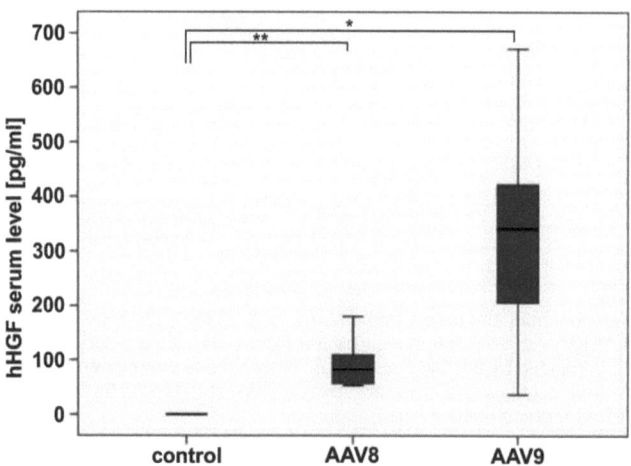

Fig. 3.2.7 Amount of hHGF determined in sera of COL4A3 knockout mice after rAAV-8 and rAAV-9 mediated recombinant hHGF expression quantified by hHGF ELISA. 5 x 10^{11} viral particles of empty capsids (control, n = 6), rAAV8-CMV-Ksp-hHGF (AAV8, n = 8) and rAAV9-CMV-Ksp-hHGF (AAV9, n = 8) were injected into 4 week old male COL4A3 knockout mice. Mice were sacrificed 5.5 weeks post injection and blood was taken for serum analyses. The hHGF amount was measured in duplicates. Statistical significance is indicated by asterisks (* = $p < 0.05$ and ** = $p < 0.001$).

3.2.4 Anti-fibrotic function of AAV-mediated hHGF expression

Since the *in vitro* experiments in this study have demonstrated that hHGF can act as an anti-fibrotic cytokine, kidneys of the transduced mice expressing hHGF via the Ksp-cadherin promoter enhanced by the CMV enhancer were analyzed with regard to their expression pattern of different genes involved in fibrosis.

3.2.4.1 Repression of fibrotic markers by recombinant hHGF expression

For the determination of the therapeutic effect of human HGF on different genes involved in fibrosis, the expression of collagen Iα1 (COL1A1), smooth muscle actin (SMA), PDGF receptor beta (PDGFRβ), and connective tissue growth factor (CTGF) were analyzed by real-time PCR. Collagen Iα1 is the main matrix protein accumulated during fibrosis, and a target of HGF regulation as already shown in the *in vitro* experiments. RNA was extracted from kidney tissues and cDNA was transcribed in order to quantify the expression of the different genes. The mRNA level of collagen Iα1 (COL1A1, Fig. 3.2.8 A) was 40 % reduced in rAAV8-CMV-Ksp-hHGF injected mice compared to control mice that had received only empty capsids. rAAV9 transduced mice showed around 50 % reduced COL1A1 mRNA levels. In addition, statistical calculations displayed a significant correlation between serum hHGF levels and COL1A1 expression levels with $R^2 = 0.409$ and $p < 0.01$ (Fig. 3.2.8 B) revealing the beneficial activity of iatrogenic expression of the cytokine hHGF.

Another gene that was analyzed in respect to modified expression mediated via the introduced hHGF was alpha-smooth muscle actin (SMA). An early event of tubulointerstitial fibrosis is the peritubular accumulation of myofibroblasts that express SMA and contribute to abnormal matrix production [10]. Thus, SMA expression conduces to an indicative marker of fibrosis initiation and progression. In concordance to collagen Iα1 expression, the expression level of SMA was also reduced by 20 % in rAAV8- and 47 % in rAAV9 treated mice compared to mice that had received empty capsids (Fig. 3.2.8 C). Again, SMA and hHGF expression showed a significant reverse correlation ($R^2 = 0.923$, $p = 0.003$).

The third marker assayed was PDGFRβ. This receptor has been associated with fibrotic conditions presumably by driving proliferation of interstitial myofibroblasts [154]. As seen for collagen Iα1 and SMA, HGF transgene expression also resulted in a down-regulation of the mRNA level of PDGFRβ (23 % reduction for rAAV8 and 45

% for rAAV9 injected mice; Fig. 3.2.8 E). The reduced mRNA levels also correlated with the elevated hHGF sera levels (R^2 = 0.233; p = 0.031, Fig. 3.2.8 F).

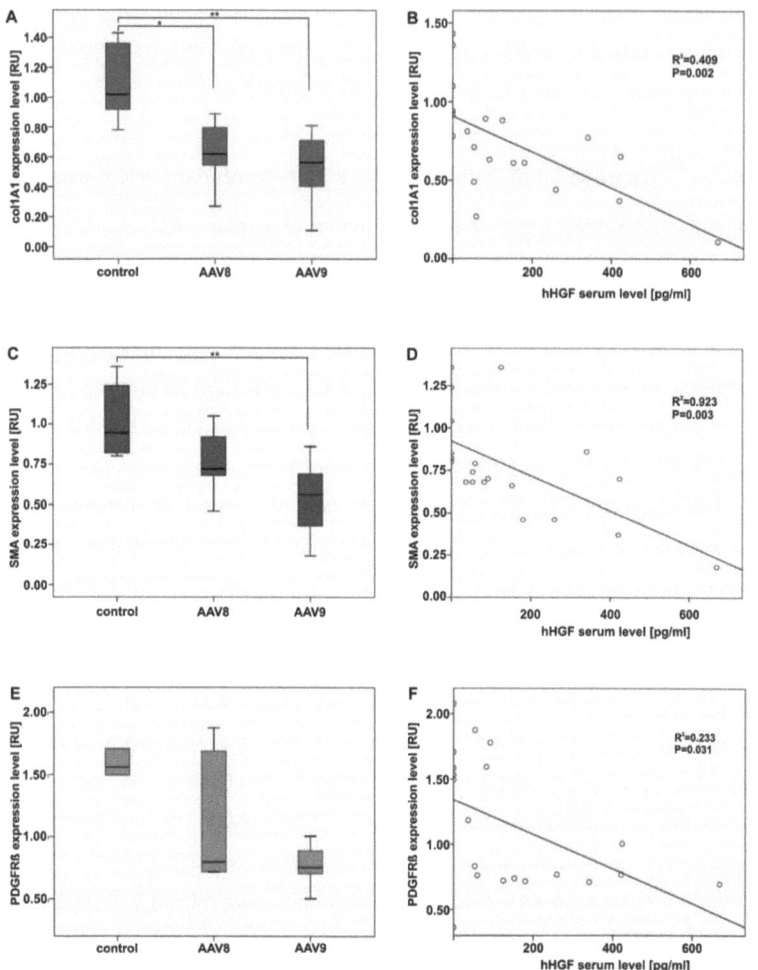

Fig. 3.2.8: Transcript levels of (A) COL1A1, (C) SMA and (E) PDGFRβ in kidneys of male COL4A3 knockout mice transduced with 5 x 10^{11} particles of empty capsids (control; n = 6), rAAV8-CMV-Ksp-hHGF (AAV8; n = 8) or rAAV9-CMV-Ksp-hHGF (AAV9; n = 8), respectively. The AAV vectors were injected at the age of 4 weeks and the mice were sacrificed 5.5 weeks post injection. The kidneys were dissected and RNA was extracted of half a kidney. The RNA was reverse transcribed and real-time PCR was performed. The expression levels were normalized to HPRT and were calculated as relative units [RU] using a standard curve. The statistic correlation between the hHGF sera level and COL1A1 (B), SMA (D) and PDGFRβ (F) is shown by linear regression using SPSS. Significance is indicated by asterisks for A, C and E (* = p < 0.05; ** p < 0.01) or given as p-value (for B, D, F).

Further, the influence of hHGF on the expression pattern of connective tissue growth factor (CTGF) was investigated by real-time PCR. CTGF is a downstream mediator for TGFβ that has been shown to up-regulate the production of extracellular matrix proteins [162, 163] and has been linked to chronic tubulointerstitial fibrosis [162]. As displayed in figure 3.2.9, there is no difference in the CTGF mRNA levels between the rAAV8 and rAAV9 hHGF treated and untreated mice detectable, since they exhibit nearly the same expression levels.

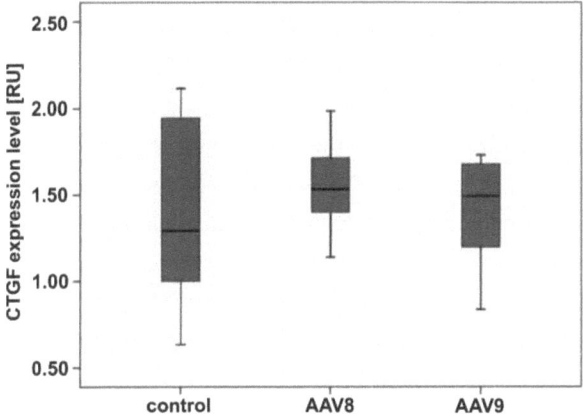

Fig. 3.2.9: Transcript levels of CTGF in male COL4A3 knockout mice transduced with empty capsids (control; n = 6), rAAV8-CMV-Ksp-hHGF (AAV8; n = 8) or rAAV9-CMV-Ksp-hHGF (AAV9; n = 8). 5×10^{11} viral particles of the AAV vectors were injected at the age of 4 weeks and the mice were sacrificed 5.5 weeks post injection. The kidneys were dissected and RNA was extracted of half a kidney. The RNA was reverse transcribed and real-time PCR was performed. The expression levels were normalized to HPRT and were calculated as relative units [RU] using a standard curve.

In summary, the results of the experiments demonstrate that the expression levels of the three master genes involved in fibrotic processes were highly down-regulated by exogenous hHGF expression. The most efficient anti-fibrotic effect was obvious by the delivery of hHGF via rAAV9. By the use of this vector as gene vehicle the highest hHGF serum level was achieved, which was accompanied by the most efficient down-regulation of the mRNA level of COL1A1, SMA and PDGFRβ.

In addition to the investigations of the effect of hHGF on the mRNA level of fibrotic genes, the fibrotic alterations in the kidney were studied by histological monitoring.

3.2.3.6 Deceleration of fibrotic remodelling of the kidney architecture

Renal interstitial fibrosis is characterized by an increase of extracellular matrix (ECM) deposition. Therefore, accumulation of ECM was examined by histological gomori staining that recognizes reticulin fibers of the connective tissues. The classification of the different stages was carried out as described in 2.2.10.5. As shown in figure 3.2.10 A, treatment with hHGF significantly attenuated interstitial accumulation of connective tissue. While five out of six mice that received empty capsids displayed a fibrosis grade of 4 and 5 (Fig. 3.2.10 B) in the kidney, six out of 8 rAAV8-CMV-Ksp-hHGF injected mice revealed with a fibrosis grade of 3 and thus a reduced renal extracellular matrix deposition compared to control mice. Actually, the rAAV9-CMV-Ksp-hHGF transduced mice showed the strongest effect. Four out of eight rAAV9-CMV-Ksp-hHGF injected mice showed a fibrosis grade of 2 in the kidneys. Three mice were classified with a grade of 3 and one mouse exhibited renal extracellular matrix accumulation with a grade of 4. As shown in figure 3.2.10 C the stage of fibrosis was determined visually by light microscopy. In figure 3.2.10. (A) the different stages are shown. The interstitial accumulation of connective tissue was mainly detectable in the medulla region, apparent as brownish staining. A linear correlation of the fibrotic progress and the hHGF sera level displayed significance with $p < 0.01$.

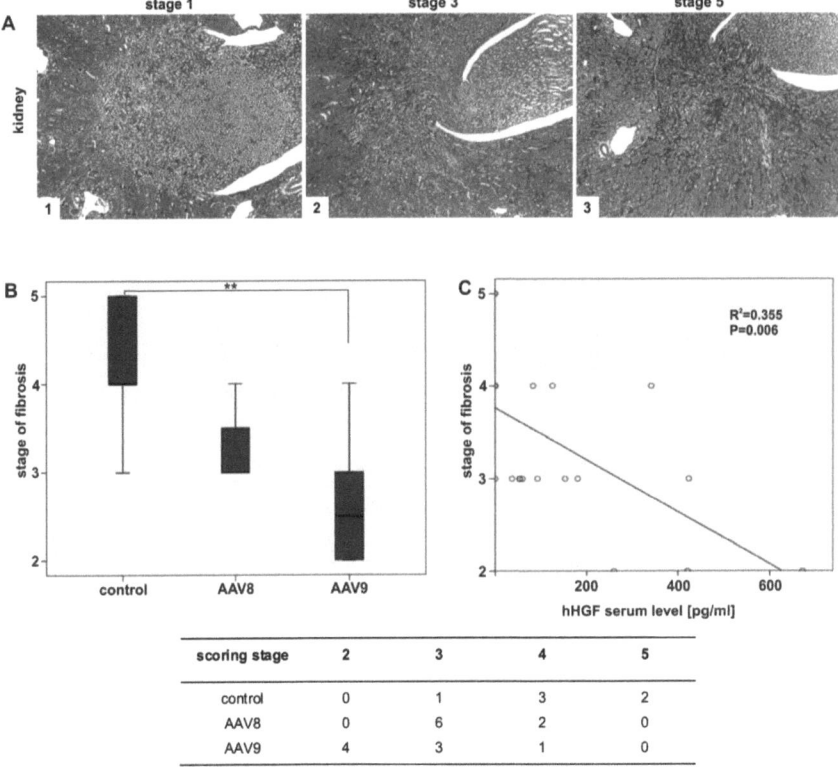

Fig. 3.2.10: Staining and staging of extracellular matrix deposition in transversal kidney sections of male COL4A3 knockout mice after rAAV8 and rAAV9 mediated recombinant hHGF expression by gomori staining.
(A) Gomori staining of transversal sections of the kidneys with fibrosis representing stages 1, 3, and 5 according to scoring described in material and methods (scale bar = 40 µm).
(B) Boxplott displaying the reduction of connective tissue deposition in the rAAV8-CMV-Ksp-hHGF (AAV8; n = 8) and rAAV9-CMV-Ksp-hHGF (AAV9; n = 8) treated mice compared to the control mice (control; n = 6) (** = $p < 0.01$), summarized also in the table.
(C) Correlation of the gomori stages and the hHGF sera levels. $p < 0.05$ were considered significant.

4. Discussion

HGF is a central growth factor that mediates not only regeneration but also anti-fibrotic processes. However, little is known about the underlying molecular and cellular mechanisms of the anti-fibrotic actions by HGF. In the present study, profiling of anti-fibrotic HGF-initiated signal transduction of Erk1/2 and Akt revealed for the first time that in addition to the fibrotic markers collagen 1A1 and SMA a wide panel of fibrotic mediators are dysregulated. New HGF targets of anti-fibrotic action were not only identified *in vitro*, but also *in vivo* such as PDGF receptor beta (PDGFRβ) and members of the CCN family. Tubulointerstitial fibrosis has been shown to be a histological predictor for end stage renal failure. Renal interstitial fibroblasts represent an important cell type in this process as they are involved in the secretion of extracellular matrix markers and mediators. HGF targeting by various AAV vectors demonstrates that rAAV9 combined with the Ksp promoter is ideal to deliver HGF to renal fibrotic lesions upon tubular injury and tubulointerstitial fibrosis in a COL4A3 knockout model resembling the human Alport disease.

4.1 HGF mediates its anti-fibrotic effects by MAP/Erk- and by Akt-signaling

In order to gain insight into the signal transduction of HGF in interstitial fibrosis, the phosphorylation statuses of central transducers of HGF signaling, Erk1/2, Akt and Stat3, were analyzed in renal interstitial fibroblasts. These analyses revealed that in agreement to data collected in epithelial cells [29, 164], HGF caused not only Erk1/2 but also Akt activation. Furthermore, it turned out that in response to HGF induced Erk1/2 activation, Smad2 was phosphorylated at the linker region, ten minutes later. These observations are consistent with previous data of Yang et al. who demonstrated Smad2/3 phosphorylation in dependence of active Erk1/2 [68]. They postulated that activated Erk1/2 is responsible for the phosphorylation of the linker region of Smad2, containing three Ser-Pro sequences (Ser-245, Ser-250, Ser-255) serving as potential phosphorylation sites for Erk1/2 [29]. This, in turn, impedes Smad translocation into the nucleus, thereby abolishing TGFβ mediated responses [29, 68]. In epithelial cells, which constitutively present the c-met receptor thereby mediating mitogenic and morphogenic HGF signals in terms of homeostasis and regeneration processes, interaction of HGF with the Smad transcriptional co-repressor SnoN is also described [152]. In contrast to inhibited nuclear translocation of activated Smad2/3 by HGF in fibroblasts, SnoN, once up-regulated by HGF in epithelial cells, binds to activated Smads in the nucleus thereby blocking gene transcription [152]. A similar mechanism is reported for mesangial cells. Instead of SnoN another Smad-corepressor, TGIF, is induced by HGF and sequesters the transcriptional activation of TGFβ target genes [69].

Beside activation of the MAP/Erk pathway, already proven by Yang and his colleagues [68], this study demonstrates that a second signaling pathway turned on by HGF in renal fibroblasts: the Akt pathway. The onset of this signaling cascade is simultaneous to the activation of the MAP/Erk pathway and HGF results in a phosphorylation of Akt, visible already five minutes after stimulation. Akt (protein kinase B) is one of the major downstream targets of phosphoinositol-3 kinase (PI3K) [165] and has been implicated in the regulation of multiple cellular functions including cell growth, survival, metabolism, protein synthesis, anti-apoptosis, tumor growth, and angiogenesis [165, 166]. Especially in human cancer Akt plays a decisive role. Since the importance of HGF signaling via Akt in renal fibroblasts is not yet known, here the question arose if there is also a function of Akt in regard to fibrotic processes. Is the blockade of the Smad signaling via MAP/Erk alone responsible for the anti-fibrotic effect of HGF in interstitial fibroblasts or is the Akt signaling also involved in the mediation of anti-fibrotic effects? The knockdown of both, Smad4 and Akt, respectively, revealed that not only the Erk1/2 / Smad2 interaction is involved in down-regulation of fibrotic mediators but also the HGF stimulated Akt pathway, which will be discussed later on.

In contrast to the activation of Erk1/2 and Akt, Stat3 was not phosphorylated by HGF in renal fibroblasts. While HGF is reported to induce Stat3 phosphorylation and nuclear translocation in MDCK epithelial cells inducing tubulogenesis [167] and in mesangial cells, supporting proliferation [168] the Stat3 pathway in interstitial fibroblasts is not turned on by HGF.

4.2 Profiling of anti-fibrotic signals in interstitial fibroblasts

In order to analyze molecular implications of HGF initiated Erk1/2 and Akt signaling in renal fibroblasts, a comprehensive expression profiling was performed. The screening displayed numerous genes to be dramatically affected by HGF. The up-regulated genes were preferentially involved in cellular motor activity and intracellular transporter activity. Mainly, the expression of kinesins and myosins was increased as well as the expression of nucleoporins. In order to identify HGF affected genes that might be involved in fibrogenesis, the focus was attached to candidates selected by their pronounced divergent expression profile and their potential association with pro-fibrotic processes. The expression level of collagen type I (COL1A1) and SMA that are strongly linked to fibrosis were highly down-regulated upon HGF stimulation of renal fibroblasts [58] as shown recently also in dermal and lung fibroblasts [169-171]. SMA is an indicative marker for myofibroblastic cells, which are primarily responsible for the overproduction and deposition of ECM [10, 58]. The intense decrease of SMA transcript levels by HGF support the data of Yang et al., who demonstrated a reduction of SMA in interstitial fibroblasts after HGF treatment *in vitro* [68]. Furthermore, TGFβ-induced SMA expression in epithelial cells is also shown to be inhibited by HGF *in vitro* and *in vivo* [57-59, 78]. HGF seems to inhibit the activation of interstitial fibroblasts into ECM-producing myofibroblasts, thereby mediating a reduced ECM production and deposition. In addition to SMA the expression profiling in this study identified various collagens that displayed repressed transcript levels after HGF treatment. The validation by further expression analysis demonstrated up to 70 % minimized mRNA levels of collagen type I and collagen type IV, depending on the collagen type and polypeptide chain. These data are concordant to previous reports showing that HGF resulted in a down-regulation of COL1A1, the main fibrous collagen, representing approximately 84 % of the collagen synthesized by fibroblasts [172] *in vitro* [68] and *in vivo* [57, 59, 78, 80]. Type I collagen is made up of two polypeptide chains, COL1A1 and COL1A2 [172]. Even though both are regulated by TGFβ [19, 173, 174], there is a difference in the induction of expression. While COL1A2 is reported to be regulated by TGFβ signaling via the Smad pathway [172] due to a SP-1 [172, 175] and an AP-1 [172, 176] binding site in the promoter, COL1A1 can be directly activated by TGFβ in a Smad independent manner. This activation can be attributed to an additional TGFβ response element in the distal promoter [177]. Upon TGFβ treatment a TGFβ activator protein directly binds to the TGFβ element thereby activating COL1A1 gene transcription [173, 177] without Smad activation. Verrechia and colleagues, however, reported that the COL1A1 promoter is also a Smad target [18]. Interestingly, the siRNA Smad4 knockdown resulted in a down-regulation of COL1A1 that was not exceeded by HGF, indicating a predominant Smad-dependent regulation in interstitial fibroblasts. Therefore, both polypeptide chains, COL1A1 and COL1A2, seem to be down-regulated by HGF via the activation of Erk1/2 and the following blockade of the Smad2/3 transducers. Though, another possible explanation may be due to an indirect down-regulation by decreased CTGF, the connective tissue growth factor. Several studies demonstrate that CTGF is also responsible for the induction of collagen synthesis by TGFβ [178, 179] in particular for COL1A1. Here in this study, HGF dramatically diminished CTGF expression which would explain a correlation with the reduced collagen expression. CTGF is a well accepted fibrotic mediator [180]. It belongs to the CCN family and is considered to be a down-stream regulator of the pro-fibrotic actions of TGFβ [181]. The biological activities of CTGF include stimulation of cell proliferation, DNA synthesis, survival, ECM production, and angiogenesis [163]. *In vivo* studies have shown that CTGF is overexpressed in fibrotic lesions of major organs and tissues like liver [182, 183], lung [184, 185], and kidney [186, 187]. Furthermore, its overproduction has also been linked to chronic tubulointerstitial fibrosis [162, 188, 189]. The promotor region contains multiple potential regulatory elements like AP-1 sites, SP-1 sites and a CATbox [190] as well as a functional Smad binding site for Smad3/4 [191]. An additional TGFβ response element in the promoter region is postulated to be exclusively responsible for a direct induction of CTGF by TGFβ [190]. *In vitro* approaches demonstrated that the induction of CTGF by TGFβ is restricted to fibroblasts and mesangial cells and does not occur in epithelial cells [190, 192]. In both cell types the activation of the CTGF promoter by TGFβ is reported to be cooperatively mediated via two signaling pathways, the Smad-signaling and the Ras/MAPK/Erk signaling [32, 193]. Notably, the Smad4 siRNA approach of the here presented study lead to the assumption that CTGF in interstitial fibroblasts is almost completely regulated via the Smad signaling pathway because the knockdown of Smad4 resulted in a dramatic decrease of the CTGF mRNA level. Surprisingly, the treatment of Smad4 silenced renal fibroblasts with HGF resulted in an additional negative effect that can be attributed to the Akt-pathway, as the inhibition of this signaling cascade depicted an immense increase of CTGF expression. Therefore, the regulation of CTGF seems to be under the control of the Smad-pathway as well as the Akt-pathway and furthermore the reduced expression of COL1A1 and COL1A2 by HGF is not only directly mediated but probably also due to in an indirect regulation via CTGF.

Likewise to collagen I, data supplied by expression profiling displayed decreased expression levels of COL4A1 and COL4A5 by HGF treatment. Collagen IV is a nonfibrillar protein representing the most abundant collagen type in the basal membrane of the glomerulus and the renal tubules [194]. To date there are only few analyzes with regard to the interaction of HGF and collagen IV. A reduced expression of collagen type IV by HGF is already reported for human glomeruli [195]. Additionally, glucose treated mesangial cells revealed reduced collagen IV protein levels after HGF treatment [72]. In this study, silencing of Smad4 lead to a strongly decreased COL4A5 expression, indicating a Smad-dependent regulatory mechanism. But the question, if HGF directly reduces collagen IV or indirectly still has to be solved. However, the expression of collagen IV could also be under the control of CTGF, such as collagen I. A correlation between both is reported by Liu and his coworkers. They demonstrated an increased deposition of collagen IV in tubulointerstitial fibrosis and glomerulosclerosis associated with an over-expression of CTGF [196].

In conclusion, important fibrotic markers, the collagens, are down-regulated by HGF predominantly via the blockade of the Smad-pathway. But in addition, there seems to be an auxiliary indirect regulation of the collagens by CTGF, the main fibrotic mediator down-stream of TGFβ. The reduced CTGF expression level after HGF stimulation is probably also a cause for the diminished transcript levels of the collagens.

In addition to CTGF, the expression profiling displayed a dramatic down-regulation of two other members of the CCN-family by HGF, Wisp2 (wnt-induced secreted protein-2) and Nov (nephroblastoma overexpressed protein). Especially, Nov revealed a reduced expression of more than 90 %. Like CTGF, both are also extracellular matrix associated proteins [157, 197] that are involved in stimulation of mitosis, adhesion, apoptosis, ECM production, growth arrest and migration [198]. Furthermore, they regulate angiogenesis and tumor growth [198]. While CTGF is known to play a critical role in injury repair and fibrotic diseases as already discussed above, until now little is known about the role of Wisp2 or Nov in fibrotic processes and their possible regulation by HGF. An increased expression of Nov in association with proceeding liver fibrosis is reported by Lee et al. [199]. In addition, they ascribed the induced Nov expression in hepatic stellate cells to TGFβ. While another study revealed Nov as a downstream target of PDGF-B and -D signaling via the PDGFRβ (PDGF receptor beta) in human mesangial cells [200], the regulation of Nov in interstitial fibroblasts seems to be subjected to another regulatory mechanism, because HGF resulted both in a down-regulation of PDGFRβ and Nov, with a more dramatic effect on the latter. Until now nothing is known about the Nov expression in renal interstitial cells. But particularly with regard to the intense down-regulation by HGF, a crucial role for Nov in interstitial fibrosis can be assumed. Interestingly, the regulation of Nov by HGF can be ascribed to the Smad pathway as well as to the Akt pathway. While silencing of Smad4 drastically decreased Nov expression, knockdown of Akt resulted in an enormous induction of Nov transcription.

The same was obvious for PDGFRβ, a receptor strongly connected to fibrosis. HGF treatment intensely diminished PDGFRβ expression. Knockdown of Smad4 could attribute this effect to the inhibited Smad signaling. However, Akt signaling represented an additional negative effect, as the expression of PDGFRβ was markedly raised after Akt knockdown, indicating a negative regulation of PDGFRβ by HGF via both cascades, MAP/Erk as well as Akt. The PDGF family consists of four different polypeptide chains (PDGF-A, -B, -C and –D) and two tyrosine kinase receptors (PDGFRα and β) [201-203]. In the adult kidney, both receptors are highly expressed in interstitial cells [204] and their synthesis depends on external stimuli such as other cytokines and growth factors [205]. An over-expression of the polypeptide chains as well as the receptors is linked to diseases with excessive cell growth like fibrotic disorders, malignancies and arteriosclerosis [206]. PDGF signaling has been implicated in several fibrotic conditions and is assumed to play a role in driving proliferation of cells with a myofibroblastic phenotype [154]. Especially, signaling of PDGF-B and -D plays an important role in case of organ fibrosis and only the signaling via PDGFRβ seems to be the key mediator of interstitial fibroblast proliferation [207, 208]. Depletion of the receptor by HGF intercepts the pro-fibrotic action of PDGF-B and PDGF-D thereby preventing renal failure and tubulointerstitial fibrosis. This result can be linked to a study of Bessho et al. [209]. They demonstrated that HGF was able to suppress PDGF-induced proliferation of mesangial cells *in vivo* and *in vitro*, but did not significantly change PDGF expression level in glomerular cells. Therefore, the effect of HGF in mesangial cells and interstitial fibroblasts is probably mediated not by a reduction of the pro-fibrotic PDGF-B or PDGF-D but by the reduction of the receptor for both, thereby suppressing the signaling cascade.

FIGF, a gene related to the platelet-derived growth factor/vascular endothelial growth factor family, was also strongly down-regulated by HGF stimulation. FIGF (c-fos induced growth factor) is also termed VEGF-D and is known to be induced by the proto-oncogene c-fos [155]. The transcription of c-fos is induced in response to many extracellular signals amongst others growth factors. The expression profiling, however, displayed no dysregulated mRNA level of c-fos by HGF, indicating a regulation that is independent of c-fos. The knockdown of Smad4 revealed a Smad-dependent regulation, whereas the Akt pathway is not involved in the regulation.

With regard to the major or predominant cytokine involved in fibrosis, TGFβ, as well as its downstream cytoplasmic mediators Smad2, -3 and -4, only marginal effects on the expression levels were detectable after HGF application. This argues for a regulation of TGFβ driven neither by the TGFβ signaling itself nor by the Akt signaling. Both would yield in a down-regulation of TGFβ upon HGF stimulation. In contrast, the inhibitory Smad7, antagonising TGFβ initiated Smad2/3 signaling by competitive receptor type I interaction, is more than 50 % reduced after HGF treatment. Although promoter activity of Smad7 was shown to depend on Erk1/2 initiated Ap1-binding [210], Smad7 is known to be primarily transcriptionally regulated by TGFβ in a feedback loop, due to binding of Smad3 to a TGFβ responsive site. Thus, HGF mediated repression of Smad7 is assumed to arise by reason of the induced Smad2/3 blockade after HGF-initiated ERK1/2 phosphorylation.

Chronic inflammation is believed to play a critical role in the initiation and prevention of chronic kidney diseases [211, 212]. Expression analyses could detect various targeting domains of HGF in interstitial fibroblasts, amongst others a set of chemokines that were negatively affected by HGF stimulation. The most striking chemokine that was highly down-regulated by HGF was the CC-chemokine ligand 5 (Ccl5), also known as RANTES. This chemokine plays an important role in chronic kidney diseases [213, 214] and is one of the best characterized chemokines that is able to recruit many types of immune cells [211]. Giannopoulou et al. could already show that HGF suppresses IL-1β or TNF-α-induced RANTES expression in tubular epithelial cells [211]. The effect of HGF on RANTES in interstitial fibroblasts suggests that HGF exerts its potent inhibitory influence of renal inflammation also in mesenchymal cells. It is assumed that chemokines either directly or indirectly contribute to interstitial collagen deposition and fibrosis by the recruitment of macrophages [212]. Therefore the alteration of the pro-fibrotic cytokine profile in interstitial fibroblasts by HGF may also account for the anti-fibrotic effect and especially the reduced collagen synthesis.

In summary, the presented data lead to the suggestion that not only the down-regulation of collagens and CTGF are involved in the anti-fibrotic effect of HGF, but also repression of other members of the CCN family and various cytokines by HGF play a decisive role. Furthermore, it could be shown that HGF stimulation of renal fibroblasts resulted in the activation of both, the Erk1/2 and the Akt pathway and that not only the Erk1/2 / Smad2 interaction is involved in the down-regulation of fibrotic mediators but also the HGF stimulated Akt pathway.

4.3 A gene therapeutical approach targeting HGF to renal failure, *in vivo*

In order to analyze the anti-fibrotic effect of HGF also *in vivo*, a gene therapeutical approach was carried out using a mouse model for interstitial fibrosis and a systemic application of HGF. While earlier studies demonstrated the application of exogenous HGF via protein or its gene in chronic kidney diseases, in this study AAV vectors were utilized as gene vehicle for HGF. Tracing experiments using GFP as a reporter and three different rAAV serotypes (rAAV2, rAAV8 and rAAV9) as gene vehicle revealed an insufficient systemic delivery of rAAV2 to the liver and the kidney, whereas the alternate serotypes rAAV8 and rAAV9 successfully transduced both organs. The HGF targeting by rAAV8 and rAAV9 demonstrated that rAAV9 combined with the Ksp promoter was best suited for the delivery of HGF to renal fibrotic lesions upon tubular injury and tubulointerstitial fibrosis in the COL4A3 knockout model that resembles the human Alport disease. The anti-fibrotic effect of AAV-mediated HGF expression in the kidney and the liver resulted not only in a remarkable reduction of the expression of fibrotic associated genes but also in a considerable reduction of the severity of fibrosis. These data reveal a novel strategy for the treatment of chronic kidney diseases.

4.3.1 HGF applied by the adeno-associated virus *in vivo*

Previous work has already demonstrated the anti-fibrotic function of locally or systemically applied HGF in a variety of experimental systems including models of murine and porcine renal failure [57-59, 78, 80, 215]. However, these studies report of a HGF application by recombinant protein or plasmid DNA. Systemically applied HGF protein is cleared extremely fast from blood circulation (half-life 3-5 minutes) making it costly to reach therapeutic blood levels and also the local application needs repeated administration. A solution to this problem would be a gene therapeutical approach allowing for a continuous expression of HGF. Thus, the intention of this study was a gene therapy for Alport syndrome via systemic application of HGF. Since AAV vectors are non-pathogenic and low immunogenic, allow for a stable gene transfer and transduce both, dividing and non-dividing cells [124, 216, 217] they were chosen as gene-transfer vehicles. Based on the different tropisms for each serotype, the first intention was to identify an AAV vector best suited for gene delivery to the kidney, to reach a continuous expression of the anti-fibrotic transgene HGF locally.

4.3.2 AAV9 is superior to AAV8 and AAV2 in kidney and liver transduction

The here reported tracing experiments with GFP as reporter and three different serotypes (rAAV2, -8, and -9) as gene vehicles revealed an insufficient systemic delivery for rAAV2. Only low numbers of vector genomes could be detected within the kidney, and - in accordance - neither by quantitative real-time PCR nor by immunohistochemical analyses significant reporter gene expression could be detected. In comparison to the kidney, the liver contained a higher level of rAAV2 delivered vector genomes. Transgene expression, however, was extremely low despite the use of self-complementary vector genomes. Self-complementary vector genomes have the advantage to overcome the limiting necessary second strand synthesis to obtain a double stranded DNA-template for initiation of gene expression, as they are already double-stranded [218].

In contrast, the alternate serotypes rAAV8 and rAAV9 were able to transduce the liver and the kidney, reaching the highest expression rate mediated by rAAV9. Interestingly, GFP expression was primarily visible in the medulla-region, the region where interstitial fibrosis and deposition of extracellular matrix is mainly observable, respectively, indicating a local expression by AAV at the place of interest, the region of ECM accumulation. These data are concordant with other reports even if there are just few reports concerning intraveneously injected AAV and the kidney. As reported by Takeda and colleagues, local delivery of AAV2 into rats via the renal artery by catheterization successfully transduced tubular epithelia cells, but not glomerular, blood vessel, or interstitial cells, while neither of the other serotypes (rAAV-1, -3 to -5) showed any transduction [123]. Alternative approaches for local kidney gene delivery were reported by Lipkowitz and colleagues, who injected rAAV2-GFP intraparenchymally into mice kidney, thereby transducing tubular epithelial cells [141]. In addition a report of Chen et al. demonstrated successful transduction of renal tubular cells via intra-renal arterial delivery of rAAV2-GFP [219]. There are also recent data of systemic transduction of the kidney by AAV9. Although this serotype is preferentially known to be a suitable vector for cardiac transduction [120], Bostick et al. reported an efficient transduction of the kidney of adult mice by AAV9 [125]. Likewise the group of Nakai demonstrated a transduction of the kidney by rAAV9 [142].

4.3.3 Choice of a specific promoter

For the tracing experiments, the expression of the reporter GFP, that was used recently in connection with AAV *in vitro* and *in vivo* [219, 220], was placed under the control of the ubiquitously active CMV promoter. This conventional promoter is widely used with a high transcription rate without tissue specificity [221]. But beside the ubiquitous activity there is the suggestion that such viral promoters may be down-regulated *in vivo,* thereby preventing long-term expression. Moreover, based on the wide tissue distribution of AAV serotypes, the transgene expression should be restricted to the kidney. Thus, a mammalian promoter was used for the expression of HGF to be more effective and reported to be kidney-specific *in vitro* and *in vivo* [158-160]. The Ksp-promoter was selected based on the ability to limit transgene expression to the kidney, the organ of damage. To intensify the activity of the Ksp-promoter, two different enhancers were analyzed *in vitro*. While the construct containing the beta globin minimal promoter behind the Ksp-promoter demonstrated no expression in renal epithelial cells at all, the construct with the CMV enhancer in front of the Ksp-promoter reached more than 30 % activity in renal epithelial cells compared to the CMV promoter. Therefore, the Ksp-promoter strengthened by the CMV enhancer was used for further *in vivo* analyses.

However, in contrast to previous studies who demonstrated the Ksp-promoter to be exclusively expressed in tubular epithelial cells in the kidney [158], strong GFP signals in the kidney but also in the liver were detectable on transcript and protein level by *in vivo* reporter gene analyses of the Ksp-promoter enforced by the CMV enhancer. Compared to the ubiquitously active CMV promoter, the Ksp-promoter construct could demonstrate tissue restriction of the transgene expression to the liver and the kidney. While the CMV promoter mediated GFP expression in all organs studied (liver, kidney, lung, spleen and heart), the tissue specific promoter was not active in spleen, lung and heart.

Based on these *in vivo* reporter gene analyses of the tissue specific promoter construct, the gene therapeutical approach was developed as a bidirectional strategy. The liver was used as primary and kidney as secondary gene transfer target for the treatment of renal failure by rAAV mediated gene expression of the anti-fibrotic factor hHGF.

As rAAV8 and rAAV9 clearly exceeded rAAV2 in targeting liver and kidney in reporter gene studies, these vectors were used for the *in vivo* analyses of the anti-fibrotic effect of HGF in order to achieve both, endocrine and paracrine HGF delivery.

4.3.4 HGF mediates anti-fibrotic effects identified *in vitro* also *in vivo*

Systemic administration of HGF mediated by rAAV8 and rAAV9 revealed a successful transduction by both vectors. High levels of HGF were traceable 5.5 weeks after vector application in sera of rAAV8 and rAAV9 treated mice, whereas the controls were negative. The strongest HGF expression was detectable in mice treated with rAAV9-HGF vectors confirming the reporter gene analyses and indicating rAAV9 to be better suited for renal treatment.

With regard to the anti-fibrotic effect of HGF, there was a significant correlation of the HGF sera level and the transcript levels of genes associated with fibrosis. The transcript levels of the main fibrous collagen, COL1A1, were strongly reduced by HGF treatment. The same was obvious for SMA, the indicative marker for myofibroblastic cells. These beneficial effects of HGF are concordant with the data of earlier studies that also demonstrated the anti-fibrotic potential of HGF. The decrease of SMA and COL1A1 were recently reported in a mouse model of renal fibrosis after intravenous injection of naked plasmid encoding HGF [78], in a rat model of CsA-induced nephrotoxicity after HGF plasmid DNA electroporation [215], and in a UUO mouse model given recombinant HGF subcutaneously [59].

However, these reports describe the anti-fibrotic action of HGF in mice and rats that represent short-term induced renal scarring and thereby a fibrotic progress that is very un-physiological. On the contrary, in this study gene-therapeutical benefits were achieved for a periode of 5.5 weeks in the COL4A3 knockout mouse model, representing a preclinical pathophysiological model that develops interstitial fibrosis with steady progress, similar to the human Alport syndrome.

Additionally to diminished COL1A1 and SMA mRNA levels, the present study revealed significantly reduced transcript levels of the PDGF receptor β in mice that exhibited HGF in their sera. As previously shown (chapter 4.2), this receptor has been associated with fibrotic conditions presumably by driving proliferation of cells with a myofibroblastic phenotype [154]. Another assumption is reported by Kliem and coworkers who speculate that overproduction of the ligand PDGF-B as well as increased expression of its receptor PDGFRβ are responsible for attracting fibrotic cells in areas of tubulointerstitial injury [222]. Tang et al. reported that administration of exogenous PDGF-B in healthy rats caused stimulated tubulointerstitial proliferation and increased tubulointerstitial matrix accumulation [208]. A study by Taneda et al. [203] analyzed the role of PDGF-D in a mouse model of UUO. They could show that beside an increase of PDGF-B in areas of tubulointerstitial fibrosis also an increase of PDGF-D was detectable. While the latter activates preferentially PDGFRβ, PDGF-B is able to activate PDGFRα, PDGFRβ or its heterodimer PDGFRαβ [201, 202]. To date there is not much known about the interaction of HGF with PDGFRβ. Only one report describes a possible counteraction of PDGF-induced mesangial cell proliferation and a negative regulation by HGF [209]. The down-regulation of the receptor in interstitial fibrosis via HGF could mediate the delay of further expansion of myofibroblastic cells thereby preventing the scarring by deposition of extracellular matrix. The consequence is a reduced proliferation of the myofibroblastic cells going along with an inhibited accumulation of myofibroblasts in the interstitium.

DISCUSSION

Unexpectedly, connective tissue growth factor (CTGF) was not affected in a HGF-dependent manner. There was neither a significant decrease detectable in mice that received HGF, nor a correlation to the HGF sera level. As already discussed above, CTGF is a downstream mediator for TGFβ that has been linked to chronic tubulointerstitial fibrosis. HGF is known to inhibit Smad signaling in epithelial cells [152], mesangial cells [69] and fibroblasts [68]. Moreover, the activation of the CTGF promoter by TGFβ can be mediated via two different signaling pathways, either the Smad signaling or the Ras/MAPK/Erk pathway [32, 193]. Thus, CTGF repression by AAV mediated, recombinant expressed HGF was assumed. Furthermore, reduced CTGF expression after administration of HGF in a mouse model of 5/6 nephrectomy has already been shown [223]. However, 5/6 nephrectomy is a renal fibrosis model combined with prominent regeneration whereas the here addressed Alport model represents the human pathophysiology of tubulointerstitial fibrosis after proteinuria [16].

Notably, beside the reduced mRNA level of COL1A1, SMA, and PDGFRβ, a strong correlation between kidney pathology and HGF sera levels were observed. In dependency to high HGF expression levels, the severity of fibrosis was remarkably reduced. Histological analyses displayed a declined deposition of extracellular matrix attending to increased HGF levels. Again, a considerably stronger anti-fibrotic effect was detectable in mice treated with rAAV9-HGF compared to rAAV8-HGF that is due to a higher expression of hHGF by rAAV9.

In summary, the data of the *in vitro* study of the first part of this thesis could also be confirmed *in vivo*. HGF treatment of mice suffering from the Alport syndrome and finally developing an interstitial fibrosis, resulted in the down-regulation of the transcript levels of fibrotic markers like collagen 1A1 and SMA as well as the fibrotic mediator PDGFRβ. Moreover, the anti-fibrotic effect was also visible histologically by significantly reduced interstitial ECM deposition in mice that expressed high levels of HGF. However, CTGF, the main downstream mediator of TGFβ, showed no dependency to the HGF expression.

The best suited serotype for the delivery of HGF seems to be AAV9, as this vector resulted in the highest HGF expression and the most efficient anti-fibrotic effect.

4.3.5 Paracrinal and endocrinal delivery of hHGF

As HGF is known to be involved in growth, invasion and metastasis of tumors, the bidirectional expression of HGF in the kidney and the liver could be a problem. However, it is known, that endogenous HGF levels increase in response to acute injuries and diseases [49, 51, 224] and that this increase is not restricted to the damaged organ. In case of an acute renal injury, there is also an increase of HGF expression detectable in distant non-injured organs like lung, liver and spleen [51, 53-55]. The group of Miyazawa postulate that even though the HGF production is not specific for the injured organ, the activation of HGF occurs exclusively in the injured tissue [55].

Since HGF is synthesized and secreted in a biologically inactive form, a single-chain precursor [43, 46, 225], it has to be activated. This activation is mediated via specific serine proteases [55] that also seem to be only activated after injury [226] (Fig. 4.3).

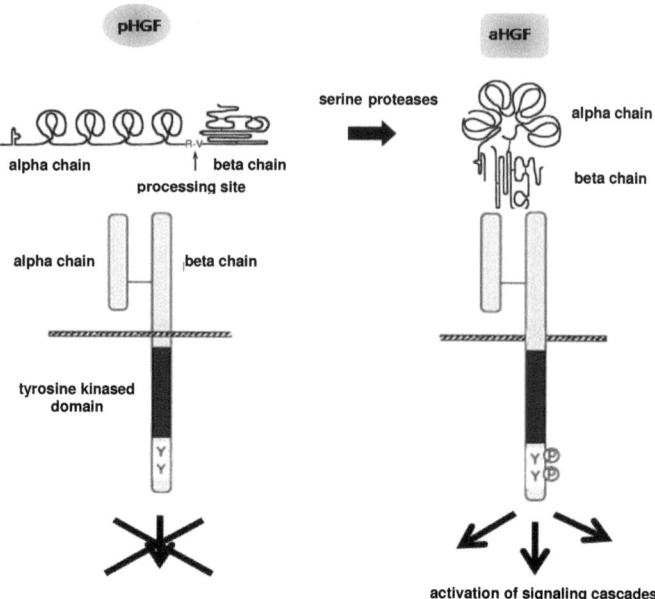

Fig. 4.3: Proteolytic conversion of HGF. HGF is synthesized and secreted in a biologically inactive form as a single-chain precursor (pHGF). This single chain precursor becomes activated by proteolytical conversion into a heavy chain and a light chain, held together by a disulfide bond (aHGF). The activation is mediated via specific serine proteases that also seem to be only activated after injury. The figure was modified according to Matsumoto et al., 2000 [47].

In addition, there are at least two reports concerning a down-regulation of the HGF receptor c-met, restricted to injured tissues [227, 228]. Endocytosis of the ligand-receptor complex implies that active HGF is only present in the damaged organ.

For the present study this activation system implicates that HGF in the circulating blood remains in an inactive single chain form unless it reaches the fibrotic kidney. Therefore, the advantage of this approach is an expression of HGF in two different organs. On one hand HGF is locally expressed in the kidney, directly in the region of injury. Notably the expression of HGF mediated by AAV occurs in the medulla, representing the highest deposition of extracellular matrix in case of interstitial fibrosis. On the other hand the AAV-mediated expression of HGF in the liver supports the anti-fibrotic effect by increased HGF levels in the circulating blood and there is no expectation of a negative effect of hHGF expressed by the liver as hHGF remains inactive until it reaches the kidney (see Fig. 4.2). A further advantage compared to earlier studies is the continuous expression of HGF.

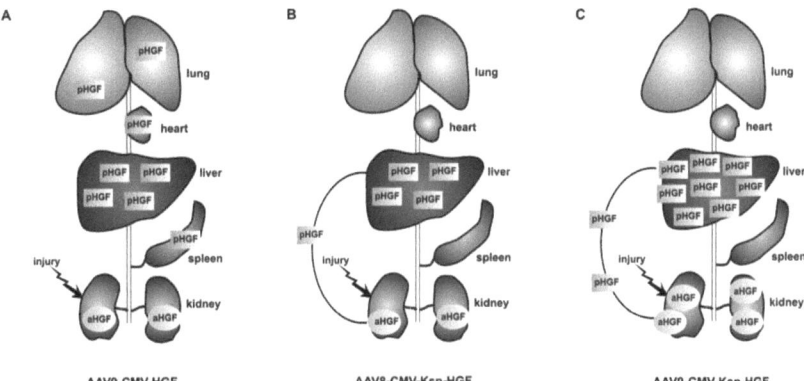

Fig. 4.2: Application of rAAV9 results in the biodistribution to a wide panel of organs and transgene expression in lung, heart, liver, spleen and kidney using the ubiquitously active CMV promoter (A). Replacement of the CMV-promoter by the Ksp-promoter enforced by the CMV enhancer restricts transgene expression to the liver and the kidney (B, C). The HGF transgene (pHGF) expressed in the liver ameliorates tubulointerstitial fibrosis after endocrine delivery and proteolytic activation (aHGF) in the kidney. Finally, application of the rAAV8-CMV-Ksp-hHGF (B) as well as rAAV9-CMV-Ksp-hHGF (C) vector provides a bidirectional therapeutical approach which is highly efficient in amelioration of chronic kidney disease by targeting efficiently both, the liver parenchyma for endocrine and the renal tubuloepithelium for paracrine transgene delivery. However, rAAV9 demonstrated to be more efficient reaching a higher hHGF expression in both organs. pHGF: pro-HGF, the inactive precursor; aHGF: activated HGF. The figure was modified according to Matsumoto et al., 2000 [47].

Taken together, this study could demonstrate for the first time a gene therapeutical approach for the treatment of interstitial fibrosis using the AAV vector as gene vehicle for the human hepatocyte growth factor. The present study took advantage on the activation mechanism at the site of injury exploiting the naive liver tropism of the AAV system to obtain an endocrinal hHGF delivery by the liver as well as a paracrinal hHGF supply in the kidney. In contrast to previous studies also using HGF as anti-fibrotic mediator this approach provides a systemic application coupled with a constant expression of HGF.

4.4 Perspectives of anti-fibrotic functions by AAV mediated HGF transfer in gene therapeutical approaches

rAAV vectors carrying HGF as transgene were constructed with the aim to mediate expression predominantly in the kidney. This was assumedly hampered by the used enhancer, the CMV enhancer, which triggered expression not only in the kidney but also in the liver. In addition, the Ksp-cadherin promoter that was reported to be kidney specific *in vitro* and *in vivo* also revealed to be slightly expressed in the liver (data not shown). Therefore, although the beneficial role of HGF in the treatment of interstitial renal fibrosis could be demonstrated and the administration via adeno-associated viral vectors revealed a novel strategy for the delivery, this system could be further improved. It would be of interest, if an expression of HGF that is exclusively restricted to the kidney is sufficient to stop the progression of interstitial fibrosis. A tissue-specific HGF expression would demonstrate if the anti-fibrotic effect shown in this study can be attributed to an HGF expression via the liver or the expression in the kidney, the place of damage. Furthermore, a transduction of only the target cell would minimize the risk of potentially negative side-effects of the transgene on other tissues. Therefore, an alternative or improved strategy would be a vector that exclusively transduces the kidney. To date there is much research on the field of AAV hybrid serotypes to enhance the efficiency of gene transfer in various tissues. As the viral capsid is responsible for the receptor binding, modifications of the capsids are under comprehensive examinations. The recently reported phage display technology [229-231] would be a promising approach to sort sequences that are exchanged in the viral capsid and exhibit desired biological properties for renal transduction.

5. References

1. Collins, A.J., et al., *Excerpts from the United States Renal Data System 2007 annual data report.* Am J Kidney Dis, 2008. **51**(1 Suppl 1): p. S1-320.
2. Xue, J.L., et al., *Forecast of the number of patients with end-stage renal disease in the United States to the year 2010.* J Am Soc Nephrol, 2001. **12**(12): p. 2753-8.
3. Wolf, G. and F.N. Ziyadeh, *Cellular and molecular mechanisms of proteinuria in diabetic nephropathy.* Nephron Physiol, 2007. **106**(2): p. p26-31.
4. Bruzzi, I., A. Benigni, and G. Remuzzi, *Role of increased glomerular protein traffic in the progression of renal failure.* Kidney Int Suppl, 1997. **62**: p. S29-31.
5. Eddy, A.A., *Experimental insights into the tubulointerstitial disease accompanying primary glomerular lesions.* J Am Soc Nephrol, 1994. **5**(6): p. 1273-87.
6. Fukagawa, M., et al., *Chronic progressive interstitial fibrosis in renal disease--are there novel pharmacological approaches?* Nephrol Dial Transplant, 1999. **14**(12): p. 2793-5.
7. Wynn, T.A., *Cellular and molecular mechanisms of fibrosis.* J Pathol, 2008. **214**(2): p. 199-210.
8. Benigni, A., *Tubulointerstitial disease mediators of injury: the role of endothelin.* Nephrol Dial Transplant, 2000. **15 Suppl 6**: p. 50-2.
9. Eddy, A.A., *Molecular insights into renal interstitial fibrosis.* J Am Soc Nephrol, 1996. **7**(12): p. 2495-508.
10. Eddy, A.A., *Molecular basis of renal fibrosis.* Pediatr Nephrol, 2000. **15**(3-4): p. 290-301.
11. Quan, T.E., S.E. Cowper, and R. Bucala, *The role of circulating fibrocytes in fibrosis.* Curr Rheumatol Rep, 2006. **8**(2): p. 145-50.
12. Liu, Y., *Epithelial to mesenchymal transition in renal fibrogenesis: pathologic significance, molecular mechanism, and therapeutic intervention.* J Am Soc Nephrol, 2004. **15**(1): p. 1-12.
13. Border, W.A. and N.A. Noble, *Transforming growth factor beta in tissue fibrosis.* N Engl J Med, 1994. **331**(19): p. 1286-92.
14. Varga, J.B., D. A.; Phan, S. H., *Fibrosis Research: Methods and Protocols.* Vol. 1. 2005: Humana Press.
15. Zeisberg, M., et al., *Stage-specific action of matrix metalloproteinases influences progressive hereditary kidney disease.* PLoS Med, 2006. **3**(4): p. e100.
16. Cosgrove, D., et al., *Collagen COL4A3 knockout: a mouse model for autosomal Alport syndrome.* Genes Dev, 1996. **10**(23): p. 2981-92.
17. Runyan, C.E., A.C. Poncelet, and H.W. Schnaper, *TGF-beta receptor-binding proteins: complex interactions.* Cell Signal, 2006. **18**(12): p. 2077-88.
18. Verrecchia, F., M.L. Chu, and A. Mauviel, *Identification of novel TGF-beta /Smad gene targets in dermal fibroblasts using a combined cDNA microarray/promoter transactivation approach.* J Biol Chem, 2001. **276**(20): p. 17058-62.
19. Yu, L., et al., *TGF-beta isoforms in renal fibrogenesis.* Kidney Int, 2003. **64**(3): p. 844-56.
20. Verrecchia, F. and A. Mauviel, *Transforming growth factor-beta and fibrosis.* World J Gastroenterol, 2007. **13**(22): p. 3056-62.
21. Saharinen, J., et al., *Latent transforming growth factor-beta binding proteins (LTBPs)--structural extracellular matrix proteins for targeting TGF-beta action.* Cytokine Growth Factor Rev, 1999. **10**(2): p. 99-117.
22. Wang, W., V. Koka, and H.Y. Lan, *Transforming growth factor-beta and Smad signalling in kidney diseases.* Nephrology (Carlton), 2005. **10**(1): p. 48-56.
23. Zhang, Y.E., *Non-Smad pathways in TGF-beta signaling.* Cell Res, 2009. **19**(1): p. 128-39.
24. Liao, J.H., et al., *The involvement of p38 MAPK in transforming growth factor beta1-induced apoptosis in murine hepatocytes.* Cell Res, 2001. **11**(2): p. 89-94.
25. Yamashita, M., et al., *TRAF6 mediates Smad-independent activation of JNK and p38 by TGF-beta.* Mol Cell, 2008. **31**(6): p. 918-24.

REFERENCES

26. Wilkes, M.C., et al., *Transforming growth factor-beta activation of phosphatidylinositol 3-kinase is independent of Smad2 and Smad3 and regulates fibroblast responses via p21-activated kinase-2.* Cancer Res, 2005. **65**(22): p. 10431-40.
27. Shi, Y. and J. Massague, *Mechanisms of TGF-beta signaling from cell membrane to the nucleus.* Cell, 2003. **113**(6): p. 685-700.
28. Macias-Silva, M., et al., *MADR2 is a substrate of the TGFbeta receptor and its phosphorylation is required for nuclear accumulation and signaling.* Cell, 1996. **87**(7): p. 1215-24.
29. Kretzschmar, M., et al., *A mechanism of repression of TGFbeta/ Smad signaling by oncogenic Ras.* Genes Dev, 1999. **13**(7): p. 804-16.
30. Derynck, R., Y. Zhang, and X.H. Feng, *Smads: transcriptional activators of TGF-beta responses.* Cell, 1998. **95**(6): p. 737-40.
31. Massague, J. and D. Wotton, *Transcriptional control by the TGF-beta/Smad signaling system.* EMBO J, 2000. **19**(8): p. 1745-54.
32. Leivonen, S.K., et al., *Smad3 and extracellular signal-regulated kinase 1/2 coordinately mediate transforming growth factor-beta-induced expression of connective tissue growth factor in human fibroblasts.* J Invest Dermatol, 2005. **124**(6): p. 1162-9.
33. Nakao, A., et al., *Identification of Smad7, a TGFbeta-inducible antagonist of TGF-beta signalling.* Nature, 1997. **389**(6651): p. 631-5.
34. Hayashi, H., et al., *The MAD-related protein Smad7 associates with the TGFbeta receptor and functions as an antagonist of TGFbeta signaling.* Cell, 1997. **89**(7): p. 1165-73.
35. Ebisawa, T., et al., *Smurf1 interacts with transforming growth factor-beta type I receptor through Smad7 and induces receptor degradation.* J Biol Chem, 2001. **276**(16): p. 12477-80.
36. Massague, J., *TGF-beta signal transduction.* Annu Rev Biochem, 1998. **67**: p. 753-91.
37. Mori, S., et al., *TGF-beta and HGF transmit the signals through JNK-dependent Smad2/3 phosphorylation at the linker regions.* Oncogene, 2004. **23**(44): p. 7416-29.
38. Shi, Y., et al., *Crystal structure of a Smad MH1 domain bound to DNA: insights on DNA binding in TGF-beta signaling.* Cell, 1998. **94**(5): p. 585-94.
39. Shi, Y., et al., *A structural basis for mutational inactivation of the tumour suppressor Smad4.* Nature, 1997. **388**(6637): p. 87-93.
40. Liu, Y., *Hepatocyte growth factor and the kidney.* Curr Opin Nephrol Hypertens, 2002. **11**(1): p. 23-30.
41. Matsumoto, K. and T. Nakamura, *Hepatocyte growth factor: renotropic role and potential therapeutics for renal diseases.* Kidney Int, 2001. **59**(6): p. 2023-38.
42. Michalopoulos, G.K. and M.C. DeFrances, *Liver regeneration.* Science, 1997. **276**(5309): p. 60-6.
43. Naka, D., et al., *Activation of hepatocyte growth factor by proteolytic conversion of a single chain form to a heterodimer.* J Biol Chem, 1992. **267**(28): p. 20114-9.
44. Naldini, L., et al., *Extracellular proteolytic cleavage by urokinase is required for activation of hepatocyte growth factor/scatter factor.* EMBO J, 1992. **11**(13): p. 4825-33.
45. Mars, W.M., R. Zarnegar, and G.K. Michalopoulos, *Activation of hepatocyte growth factor by the plasminogen activators uPA and tPA.* Am J Pathol, 1993. **143**(3): p. 949-58.
46. Miyazawa, K., et al., *Molecular cloning and sequence analysis of cDNA for human hepatocyte growth factor.* Biochem Biophys Res Commun, 1989. **163**(2): p. 967-73.
47. Matsumoto, K., S. Mizuno, and T. Nakamura, *Hepatocyte growth factor in renal regeneration, renal disease and potential therapeutics.* Curr Opin Nephrol Hypertens, 2000. **9**(4): p. 395-402.
48. Matsumoto, K. and T. Nakamura, *Hepatocyte growth factor (HGF) as a tissue organizer for organogenesis and regeneration.* Biochem Biophys Res Commun, 1997. **239**(3): p. 639-44.

49. Igawa, T., et al., *Hepatocyte growth factor may function as a renotropic factor for regeneration in rats with acute renal injury.* Am J Physiol, 1993. **265**(1 Pt 2): p. F61-9.
50. Yano, T., et al., *Regenerative response in acute renal failure due to vitamin E deficiency and glutathione depletion in rats.* Biochem Pharmacol, 1998. **56**(4): p. 543-6.
51. Liu, Y., et al., *Up-regulation of hepatocyte growth factor receptor: an amplification and targeting mechanism for hepatocyte growth factor action in acute renal failure.* Kidney Int, 1999. **55**(2): p. 442-53.
52. Oka, A., et al., *Expression of growth factors after the release of ureteral obstruction in the rat kidney.* Int J Urol, 1999. **6**(12): p. 607-15.
53. Yanagita, K., et al., *Lung may have an endocrine function producing hepatocyte growth factor in response to injury of distal organs.* Biochem Biophys Res Commun, 1992. **182**(2): p. 802-9.
54. Kono, S., et al., *Marked induction of hepatocyte growth factor mRNA in intact kidney and spleen in response to injury of distant organs.* Biochem Biophys Res Commun, 1992. **186**(2): p. 991-8.
55. Miyazawa, K., et al., *Proteolytic activation of hepatocyte growth factor in response to tissue injury.* J Biol Chem, 1994. **269**(12): p. 8966-70.
56. Ueki, T., et al., *Hepatocyte growth factor gene therapy of liver cirrhosis in rats.* Nat Med, 1999. **5**(2): p. 226-30.
57. Yang, J. and Y. Liu, *Blockage of tubular epithelial to myofibroblast transition by hepatocyte growth factor prevents renal interstitial fibrosis.* J Am Soc Nephrol, 2002. **13**(1): p. 96-107.
58. Yang, J. and Y. Liu, *Delayed administration of hepatocyte growth factor reduces renal fibrosis in obstructive nephropathy.* Am J Physiol Renal Physiol, 2003. **284**(2): p. F349-57.
59. Mizuno, S., K. Matsumoto, and T. Nakamura, *Hepatocyte growth factor suppresses interstitial fibrosis in a mouse model of obstructive nephropathy.* Kidney Int, 2001. **59**(4): p. 1304-14.
60. Bottaro, D.P., et al., *Identification of the hepatocyte growth factor receptor as the c-met proto-oncogene product.* Science, 1991. **251**(4995): p. 802-4.
61. Naldini, L., et al., *Scatter factor and hepatocyte growth factor are indistinguishable ligands for the MET receptor.* EMBO J, 1991. **10**(10): p. 2867-78.
62. Birchmeier, C. and E. Gherardi, *Developmental roles of HGF/SF and its receptor, the c-Met tyrosine kinase.* Trends Cell Biol, 1998. **8**(10): p. 404-10.
63. Comoglio, P.M. and C. Boccaccio, *The HGF receptor family: unconventional signal transducers for invasive cell growth.* Genes Cells, 1996. **1**(4): p. 347-54.
64. Graziani, A., et al., *Hepatocyte growth factor/scatter factor stimulates the Ras-guanine nucleotide exchanger.* J Biol Chem, 1993. **268**(13): p. 9165-8.
65. Ponzetto, C., et al., *A multifunctional docking site mediates signaling and transformation by the hepatocyte growth factor/scatter factor receptor family.* Cell, 1994. **77**(2): p. 261-71.
66. Matsumoto, K., T. Nakamura, and R.H. Kramer, *Hepatocyte growth factor/scatter factor induces tyrosine phosphorylation of focal adhesion kinase (p125FAK) and promotes migration and invasion by oral squamous cell carcinoma cells.* J Biol Chem, 1994. **269**(50): p. 31807-13.
67. Leshem, Y., et al., *Preferential binding of Grb2 or phosphatidylinositol 3-kinase to the met receptor has opposite effects on HGF-induced myoblast proliferation.* Exp Cell Res, 2002. **274**(2): p. 288-98.
68. Yang, J., C. Dai, and Y. Liu, *Hepatocyte growth factor suppresses renal interstitial myofibroblast activation and intercepts Smad signal transduction.* Am J Pathol, 2003. **163**(2): p. 621-32.
69. Dai, C. and Y. Liu, *Hepatocyte growth factor antagonizes the profibrotic action of TGF-beta1 in mesangial cells by stabilizing Smad transcriptional corepressor TGIF.* J Am Soc Nephrol, 2004. **15**(6): p. 1402-12.

REFERENCES

70. Mizuno, S., et al., *Reciprocal balance of hepatocyte growth factor and transforming growth factor-beta 1 in renal fibrosis in mice.* Kidney Int, 2000. **57**(3): p. 937-48.
71. Florquin, S. and K.M. Rouschop, *Reciprocal functions of hepatocyte growth factor and transforming growth factor-beta1 in the progression of renal diseases: a role for CD44?* Kidney Int Suppl, 2003(86): p. S15-20.
72. Mizuno, S. and T. Nakamura, *Suppressions of chronic glomerular injuries and TGF-beta 1 production by HGF in attenuation of murine diabetic nephropathy.* Am J Physiol Renal Physiol, 2004. **286**(1): p. F134-43.
73. Liu, Y., et al., *Endogenous hepatocyte growth factor ameliorates chronic renal injury by activating matrix degradation pathways.* Kidney Int, 2000. **58**(5): p. 2028-43.
74. Liu, Y., *Hepatocyte growth factor in kidney fibrosis: therapeutic potential and mechanisms of action.* Am J Physiol Renal Physiol, 2004. **287**(1): p. F7-16.
75. Mizuno, S., et al., *Hepatocyte growth factor prevents renal fibrosis and dysfunction in a mouse model of chronic renal disease.* J Clin Invest, 1998. **101**(9): p. 1827-34.
76. Takada, S., et al., *Effect of hepatocyte growth factor on tacrolimus-induced nephrotoxicity in spontaneously hypertensive rats.* Transpl Int, 1999. **12**(1): p. 27-32.
77. Azuma, H., et al., *Hepatocyte growth factor prevents the development of chronic allograft nephropathy in rats.* J Am Soc Nephrol, 2001. **12**(6): p. 1280-92.
78. Yang, J., C. Dai, and Y. Liu, *Systemic administration of naked plasmid encoding hepatocyte growth factor ameliorates chronic renal fibrosis in mice.* Gene Ther, 2001. **8**(19): p. 1470-9.
79. Isaka, Y., et al., *Electroporation-mediated HGF gene transfection protected the kidney against graft injury.* Gene Ther, 2005. **12**(10): p. 815-20.
80. Dai, C., et al., *Intravenous administration of hepatocyte growth factor gene ameliorates diabetic nephropathy in mice.* J Am Soc Nephrol, 2004. **15**(10): p. 2637-47.
81. Gill, D.R., I.A. Pringle, and S.C. Hyde, *Progress and prospects: the design and production of plasmid vectors.* Gene Ther, 2009. **16**(2): p. 165-71.
82. Robbins, P.D., H. Tahara, and S.C. Ghivizzani, *Viral vectors for gene therapy.* Trends Biotechnol, 1998. **16**(1): p. 35-40.
83. Gardlik, R., et al., *Vectors and delivery systems in gene therapy.* Med Sci Monit, 2005. **11**(4): p. RA110-21.
84. Langer, J.C., et al., *Adeno-associated virus gene transfer into renal cells: potential for in vivo gene delivery.* Exp Nephrol, 1998. **6**(3): p. 189-94.
85. Thomas, C.E., A. Ehrhardt, and M.A. Kay, *Progress and problems with the use of viral vectors for gene therapy.* Nat Rev Genet, 2003. **4**(5): p. 346-58.
86. Mack, C.A., et al., *Circumvention of anti-adenovirus neutralizing immunity by administration of an adenoviral vector of an alternate serotype.* Hum Gene Ther, 1997. **8**(1): p. 99-109.
87. Grimm, D. and J.A. Kleinschmidt, *Progress in adeno-associated virus type 2 vector production: promises and prospects for clinical use.* Hum Gene Ther, 1999. **10**(15): p. 2445-50.
88. Hallek, M. and C.M. Wendtner, *Recombinant adeno-associated virus (rAAV) vectors for somatic gene therapy: recent advances and potential clinical applications.* Cytokines Mol Ther, 1996. **2**(2): p. 69-79.
89. Monahan, P.E. and R.J. Samulski, *AAV vectors: is clinical success on the horizon?* Gene Ther, 2000. **7**(1): p. 24-30.
90. Wendtner, C.M., et al., *Efficient gene transfer of CD40 ligand into primary B-CLL cells using recombinant adeno-associated virus (rAAV) vectors.* Blood, 2002. **100**(5): p. 1655-61.
91. Rabinowitz, J.E. and J. Samulski, *Adeno-associated virus expression systems for gene transfer.* Curr Opin Biotechnol, 1998. **9**(5): p. 470-5.
92. Kwon, I. and D.V. Schaffer, *Designer gene delivery vectors: molecular engineering and evolution of adeno-associated viral vectors for enhanced gene transfer.* Pharm Res, 2008. **25**(3): p. 489-99.

REFERENCES

93. Le Bec, C. and A.M. Douar, *Gene therapy progress and prospects--vectorology: design and production of expression cassettes in AAV vectors.* Gene Ther, 2006. **13**(10): p. 805-13.
94. Tal, J., *Adeno-associated virus-based vectors in gene therapy.* J Biomed Sci, 2000. **7**(4): p. 279-91.
95. Atchison, R.W., B.C. Casto, and W.M. Hammon, *Adenovirus-Associated Defective Virus Particles.* Science, 1965. **149**: p. 754-6.
96. Maheshri, N., et al., *Directed evolution of adeno-associated virus yields enhanced gene delivery vectors.* Nat Biotechnol, 2006. **24**(2): p. 198-204.
97. Vasileva, A. and R. Jessberger, *Precise hit: adeno-associated virus in gene targeting.* Nat Rev Microbiol, 2005. **3**(11): p. 837-47.
98. Samulski, R.J., *Adeno-associated virus: integration at a specific chromosomal locus.* Curr Opin Genet Dev, 1993. **3**(1): p. 74-80.
99. Collaco, R.F., X. Cao, and J.P. Trempe, *A helper virus-free packaging system for recombinant adeno-associated virus vectors.* Gene, 1999. **238**(2): p. 397-405.
100. Grieger, J.C., S. Snowdy, and R.J. Samulski, *Separate basic region motifs within the adeno-associated virus capsid proteins are essential for infectivity and assembly.* J Virol, 2006. **80**(11): p. 5199-210.
101. Wu, Z., A. Asokan, and R.J. Samulski, *Adeno-associated virus serotypes: vector toolkit for human gene therapy.* Mol Ther, 2006. **14**(3): p. 316-27.
102. Auricchio, A., et al., *Exchange of surface proteins impacts on viral vector cellular specificity and transduction characteristics: the retina as a model.* Hum Mol Genet, 2001. **10**(26): p. 3075-81.
103. Gao, G.P., et al., *High-titer adeno-associated viral vectors from a Rep/Cap cell line and hybrid shuttle virus.* Hum Gene Ther, 1998. **9**(16): p. 2353-62.
104. Gao, G.P., et al., *Novel adeno-associated viruses from rhesus monkeys as vectors for human gene therapy.* Proc Natl Acad Sci U S A, 2002. **99**(18): p. 11854-9.
105. Hoggan, M.D., N.R. Blacklow, and W.P. Rowe, *Studies of small DNA viruses found in various adenovirus preparations: physical, biological, and immunological characteristics.* Proc Natl Acad Sci U S A, 1966. **55**(6): p. 1467-74.
106. Rutledge, E.A., C.L. Halbert, and D.W. Russell, *Infectious clones and vectors derived from adeno-associated virus (AAV) serotypes other than AAV type 2.* J Virol, 1998. **72**(1): p. 309-19.
107. Gao, G., et al., *Clades of Adeno-associated viruses are widely disseminated in human tissues.* J Virol, 2004. **78**(12): p. 6381-8.
108. Parks, W.P., et al., *Physicochemical characterization of adeno-associated satellite virus type 4 and its nucleic acid.* J Virol, 1967. **1**(5): p. 980-7.
109. Bantel-Schaal, U. and H. zur Hausen, *Characterization of the DNA of a defective human parvovirus isolated from a genital site.* Virology, 1984. **134**(1): p. 52-63.
110. Mori, S., et al., *Two novel adeno-associated viruses from cynomolgus monkey: pseudotyping characterization of capsid protein.* Virology, 2004. **330**(2): p. 375-83.
111. Xiao, W., et al., *Gene therapy vectors based on adeno-associated virus type 1.* J Virol, 1999. **73**(5): p. 3994-4003.
112. Chao, H., et al., *Several log increase in therapeutic transgene delivery by distinct adeno-associated viral serotype vectors.* Mol Ther, 2000. **2**(6): p. 619-23.
113. Blankinship, M.J., et al., *Efficient transduction of skeletal muscle using vectors based on adeno-associated virus serotype 6.* Mol Ther, 2004. **10**(4): p. 671-8.
114. Handa, A., et al., *Adeno-associated virus (AAV)-3-based vectors transduce haematopoietic cells not susceptible to transduction with AAV-2-based vectors.* J Gen Virol, 2000. **81**(Pt 8): p. 2077-84.
115. Rabinowitz, J.E., et al., *Cross-packaging of a single adeno-associated virus (AAV) type 2 vector genome into multiple AAV serotypes enables transduction with broad specificity.* J Virol, 2002. **76**(2): p. 791-801.
116. Weber, M., et al., *Recombinant adeno-associated virus serotype 4 mediates unique and exclusive long-term transduction of retinal pigmented epithelium in rat, dog, and nonhuman primate after subretinal delivery.* Mol Ther, 2003. **7**(6): p. 774-81.

REFERENCES

117. Zabner, J., et al., *Adeno-associated virus type 5 (AAV5) but not AAV2 binds to the apical surfaces of airway epithelia and facilitates gene transfer.* J Virol, 2000. **74**(8): p. 3852-8.
118. Halbert, C.L., J.M. Allen, and A.D. Miller, *Adeno-associated virus type 6 (AAV6) vectors mediate efficient transduction of airway epithelial cells in mouse lungs compared to that of AAV2 vectors.* J Virol, 2001. **75**(14): p. 6615-24.
119. Davidson, B.L., et al., *Recombinant adeno-associated virus type 2, 4, and 5 vectors: transduction of variant cell types and regions in the mammalian central nervous system.* Proc Natl Acad Sci U S A, 2000. **97**(7): p. 3428-32.
120. Pacak, C.A., et al., *Recombinant adeno-associated virus serotype 9 leads to preferential cardiac transduction in vivo.* Circ Res, 2006. **99**(4): p. e3-9.
121. Thomas, C.E., et al., *Rapid uncoating of vector genomes is the key to efficient liver transduction with pseudotyped adeno-associated virus vectors.* J Virol, 2004. **78**(6): p. 3110-22.
122. Nakai, H., et al., *Unrestricted hepatocyte transduction with adeno-associated virus serotype 8 vectors in mice.* J Virol, 2005. **79**(1): p. 214-24.
123. Takeda, S., et al., *Successful gene transfer using adeno-associated virus vectors into the kidney: comparison among adeno-associated virus serotype 1-5 vectors in vitro and in vivo.* Nephron Exp Nephrol, 2004. **96**(4): p. e119-26.
124. Shimpo, M., et al., *Gene transfer into rat renal cells using adeno-associated virus vectors.* Am J Nephrol, 2000. **20**(3): p. 242-7.
125. Bostick, B., et al., *Systemic AAV-9 transduction in mice is influenced by animal age but not by the route of administration.* Gene Ther, 2007. **14**(22): p. 1605-9.
126. Akache, B., et al., *The 37/67-kilodalton laminin receptor is a receptor for adeno-associated virus serotypes 8, 2, 3, and 9.* J Virol, 2006. **80**(19): p. 9831-6.
127. Kashiwakura, Y., et al., *Hepatocyte growth factor receptor is a coreceptor for adeno-associated virus type 2 infection.* J Virol, 2005. **79**(1): p. 609-14.
128. Qing, K., et al., *Human fibroblast growth factor receptor 1 is a co-receptor for infection by adeno-associated virus 2.* Nat Med, 1999. **5**(1): p. 71-7.
129. Summerford, C., J.S. Bartlett, and R.J. Samulski, *AlphaVbeta5 integrin: a co-receptor for adeno-associated virus type 2 infection.* Nat Med, 1999. **5**(1): p. 78-82.
130. Summerford, C. and R.J. Samulski, *Membrane-associated heparan sulfate proteoglycan is a receptor for adeno-associated virus type 2 virions.* J Virol, 1998. **72**(2): p. 1438-45.
131. Blackburn, S.D., R.A. Steadman, and F.B. Johnson, *Attachment of adeno-associated virus type 3H to fibroblast growth factor receptor 1.* Arch Virol, 2006. **151**(3): p. 617-23.
132. Rabinowitz, J.E., et al., *Cross-dressing the virion: the transcapsidation of adeno-associated virus serotypes functionally defines subgroups.* J Virol, 2004. **78**(9): p. 4421-32.
133. Di Pasquale, G., et al., *Identification of PDGFR as a receptor for AAV-5 transduction.* Nat Med, 2003. **9**(10): p. 1306-12.
134. Walters, R.W., et al., *Binding of adeno-associated virus type 5 to 2,3-linked sialic acid is required for gene transfer.* J Biol Chem, 2001. **276**(23): p. 20610-6.
135. Choi, V.W., D.M. McCarty, and R.J. Samulski, *AAV hybrid serotypes: improved vectors for gene delivery.* Curr Gene Ther, 2005. **5**(3): p. 299-310.
136. Hildinger, M., et al., *Hybrid vectors based on adeno-associated virus serotypes 2 and 5 for muscle-directed gene transfer.* J Virol, 2001. **75**(13): p. 6199-203.
137. Lai, L.W., G.W. Moeckel, and Y.H. Lien, *Kidney-targeted liposome-mediated gene transfer in mice.* Gene Ther, 1997. **4**(5): p. 426-31.
138. Oberbauer, R., G.F. Schreiner, and T.W. Meyer, *Renal uptake of an 18-mer phosphorothioate oligonucleotide.* Kidney Int, 1995. **48**(4): p. 1226-32.
139. Bosch, R.J., A.S. Woolf, and L.G. Fine, *Gene transfer into the mammalian kidney: direct retrovirus-transduction of regenerating tubular epithelial cells.* Exp Nephrol, 1993. **1**(1): p. 49-54.

REFERENCES

140. Moullier, P., et al., *Adenoviral-mediated gene transfer to renal tubular cells in vivo.* Kidney Int, 1994. **45**(4): p. 1220-5.
141. Lipkowitz, M.S., et al., *Transduction of renal cells in vitro and in vivo by adeno-associated virus gene therapy vectors.* J Am Soc Nephrol, 1999. **10**(9): p. 1908-15.
142. Inagaki, K., et al., *Robust systemic transduction with AAV9 vectors in mice: efficient global cardiac gene transfer superior to that of AAV8.* Mol Ther, 2006. **14**(1): p. 45-53.
143. Wagner, J.A., et al., *Efficient and persistent gene transfer of AAV-CFTR in maxillary sinus.* Lancet, 1998. **351**(9117): p. 1702-3.
144. Wagner, J.A., et al., *Safety and biological efficacy of an adeno-associated virus vector-cystic fibrosis transmembrane regulator (AAV-CFTR) in the cystic fibrosis maxillary sinus.* Laryngoscope, 1999. **109**(2 Pt 1): p. 266-74.
145. Moss, R.B., et al., *Repeated adeno-associated virus serotype 2 aerosol-mediated cystic fibrosis transmembrane regulator gene transfer to the lungs of patients with cystic fibrosis: a multicenter, double-blind, placebo-controlled trial.* Chest, 2004. **125**(2): p. 509-21.
146. Kay, M.A., et al., *Evidence for gene transfer and expression of factor IX in haemophilia B patients treated with an AAV vector.* Nat Genet, 2000. **24**(3): p. 257-61.
147. Manno, C.S., et al., *AAV-mediated factor IX gene transfer to skeletal muscle in patients with severe hemophilia B.* Blood, 2003. **101**(8): p. 2963-72.
148. Manno, C.S., et al., *Successful transduction of liver in hemophilia by AAV-Factor IX and limitations imposed by the host immune response.* Nat Med, 2006. **12**(3): p. 342-7.
149. Gross, O., et al., *Preemptive ramipril therapy delays renal failure and reduces renal fibrosis in COL4A3-knockout mice with Alport syndrome.* Kidney Int, 2003. **63**(2): p. 438-46.
150. Peek, M., et al., *Unusual proteolytic activation of pro-hepatocyte growth factor by plasma kallikrein and coagulation factor XIa.* J Biol Chem, 2002. **277**(49): p. 47804-9.
151. Shimomura, T., et al., *Activation of hepatocyte growth factor by two homologous proteases, blood-coagulation factor XIIa and hepatocyte growth factor activator.* Eur J Biochem, 1995. **229**(1): p. 257-61.
152. Yang, J., C. Dai, and Y. Liu, *A novel mechanism by which hepatocyte growth factor blocks tubular epithelial to mesenchymal transition.* J Am Soc Nephrol, 2005. **16**(1): p. 68-78.
153. Recio, J.A. and G. Merlino, *Hepatocyte growth factor/scatter factor induces feedback up-regulation of CD44v6 in melanoma cells through Egr-1.* Cancer Res, 2003. **63**(7): p. 1576-82.
154. Andrae, J., R. Gallini, and C. Betsholtz, *Role of platelet-derived growth factors in physiology and medicine.* Genes Dev, 2008. **22**(10): p. 1276-312.
155. Orlandini, M., et al., *Identification of a c-fos-induced gene that is related to the platelet-derived growth factor/vascular endothelial growth factor family.* Proc Natl Acad Sci U S A, 1996. **93**(21): p. 11675-80.
156. Liu, Y., *Renal fibrosis: new insights into the pathogenesis and therapeutics.* Kidney Int, 2006. **69**(2): p. 213-7.
157. Chen, C.C. and L.F. Lau, *Functions and mechanisms of action of CCN matricellular proteins.* Int J Biochem Cell Biol, 2008.
158. Shao, X., et al., *A minimal Ksp-cadherin promoter linked to a green fluorescent protein reporter gene exhibits tissue-specific expression in the developing kidney and genitourinary tract.* J Am Soc Nephrol, 2002. **13**(7): p. 1824-36.
159. Whyte, D.A., et al., *Ksp-cadherin gene promoter. I. Characterization and renal epithelial cell-specific activity.* Am J Physiol, 1999. **277**(4 Pt 2): p. F587-98.
160. Igarashi, P., et al., *Ksp-cadherin gene promoter. II. Kidney-specific activity in transgenic mice.* Am J Physiol, 1999. **277**(4 Pt 2): p. F599-610.
161. Samulski, R.J., L.S. Chang, and T. Shenk, *A recombinant plasmid from which an infectious adeno-associated virus genome can be excised in vitro and its use to study viral replication.* J Virol, 1987. **61**(10): p. 3096-101.

162. Yokoi, H., et al., *Role of connective tissue growth factor in fibronectin expression and tubulointerstitial fibrosis.* Am J Physiol Renal Physiol, 2002. **282**(5): p. F933-42.
163. Brigstock, D.R., *The connective tissue growth factor/cysteine-rich 61/nephroblastoma overexpressed (CCN) family.* Endocr Rev, 1999. **20**(2): p. 189-206.
164. Gong, R., A. Rifai, and L.D. Dworkin, *Activation of PI3K-Akt-GSK3beta pathway mediates hepatocyte growth factor inhibition of RANTES expression in renal tubular epithelial cells.* Biochem Biophys Res Commun, 2005. **330**(1): p. 27-33.
165. Jiang, B.H. and L.Z. Liu, *PI3K/PTEN signaling in tumorigenesis and angiogenesis.* Biochim Biophys Acta, 2008. **1784**(1): p. 150-8.
166. Zdychova, J., et al., *Renal activity of Akt kinase in experimental Type 1 diabetes.* Physiol Res, 2008. **57**(5): p. 709-15.
167. Boccaccio, C., et al., *Induction of epithelial tubules by growth factor HGF depends on the STAT pathway.* Nature, 1998. **391**(6664): p. 285-8.
168. Kalechman, Y., et al., *Inhibition of interleukin-10 by the immunomodulator AS101 reduces mesangial cell proliferation in experimental mesangioproliferative glomerulonephritis: association with dephosphorylation of STAT3.* J Biol Chem, 2004. **279**(23): p. 24724-32.
169. Jinnin, M., et al., *Effects of hepatocyte growth factor on the expression of type I collagen and matrix metalloproteinase-1 in normal and scleroderma dermal fibroblasts.* J Invest Dermatol, 2005. **124**(2): p. 324-30.
170. Bogatkevich, G.S., et al., *Down-regulation of collagen and connective tissue growth factor expression with hepatocyte growth factor in lung fibroblasts from white scleroderma patients via two signaling pathways.* Arthritis Rheum, 2007. **56**(10): p. 3468-77.
171. Jiang, D., et al., *Suppression of the production of extracellular matrix and alpha-smooth muscle actin induced by transforming growth factor-beta1 in fibroblasts of the flexor tendon sheath by hepatocyte growth factor.* Scand J Plast Reconstr Surg Hand Surg, 2008. **42**(4): p. 169-73.
172. Cutroneo, K.R., *How is Type I procollagen synthesis regulated at the gene level during tissue fibrosis.* J Cell Biochem, 2003. **90**(1): p. 1-5.
173. Cutroneo, K.R., *TGF-beta-induced fibrosis and SMAD signaling: oligo decoys as natural therapeutics for inhibition of tissue fibrosis and scarring.* Wound Repair Regen, 2007. **15 Suppl 1**: p. S54-60.
174. Li, J.H., et al., *Smad7 inhibits fibrotic effect of TGF-Beta on renal tubular epithelial cells by blocking Smad2 activation.* J Am Soc Nephrol, 2002. **13**(6): p. 1464-72.
175. Inagaki, Y., S. Truter, and F. Ramirez, *Transforming growth factor-beta stimulates alpha 2(I) collagen gene expression through a cis-acting element that contains an Sp1-binding site.* J Biol Chem, 1994. **269**(20): p. 14828-34.
176. Chung, K.Y., et al., *An AP-1 binding sequence is essential for regulation of the human alpha2(I) collagen (COL1A2) promoter activity by transforming growth factor-beta.* J Biol Chem, 1996. **271**(6): p. 3272-8.
177. Ritzenthaler, J.D., et al., *Transforming-growth-factor-beta activation elements in the distal promoter regions of the rat alpha 1 type I collagen gene.* Biochem J, 1991. **280 (Pt 1)**: p. 157-62.
178. Frazier, K., et al., *Stimulation of fibroblast cell growth, matrix production, and granulation tissue formation by connective tissue growth factor.* J Invest Dermatol, 1996. **107**(3): p. 404-11.
179. Duncan, M.R., et al., *Connective tissue growth factor mediates transforming growth factor beta-induced collagen synthesis: down-regulation by cAMP.* FASEB J, 1999. **13**(13): p. 1774-86.
180. Gupta, S., et al., *Connective tissue growth factor: potential role in glomerulosclerosis and tubulointerstitial fibrosis.* Kidney Int, 2000. **58**(4): p. 1389-99.
181. Inoue, T., et al., *TGF-beta1 and HGF coordinately facilitate collagen turnover in subepithelial mesenchyme.* Biochem Biophys Res Commun, 2002. **297**(2): p. 255-60.
182. Paradis, V., et al., *Expression of connective tissue growth factor in experimental rat and human liver fibrosis.* Hepatology, 1999. **30**(4): p. 968-76.

183. Williams, E.J., et al., *Increased expression of connective tissue growth factor in fibrotic human liver and in activated hepatic stellate cells.* J Hepatol, 2000. **32**(5): p. 754-61.
184. Lasky, J.A., et al., *Connective tissue growth factor mRNA expression is upregulated in bleomycin-induced lung fibrosis.* Am J Physiol, 1998. **275**(2 Pt 1): p. L365-71.
185. Allen, J.T., et al., *Enhanced insulin-like growth factor binding protein-related protein 2 (Connective tissue growth factor) expression in patients with idiopathic pulmonary fibrosis and pulmonary sarcoidosis.* Am J Respir Cell Mol Biol, 1999. **21**(6): p. 693-700.
186. Goldschmeding, R., et al., *Connective tissue growth factor: just another factor in renal fibrosis?* Nephrol Dial Transplant, 2000. **15**(3): p. 296-9.
187. Clarkson, M.R., et al., *Connective tissue growth factor: a potential stimulus for glomerulosclerosis and tubulointerstitial fibrosis in progressive renal disease.* Curr Opin Nephrol Hypertens, 1999. **8**(5): p. 543-8.
188. Wang, S., et al., *Connective tissue growth factor in tubulointerstitial injury of diabetic nephropathy.* Kidney Int, 2001. **60**(1): p. 96-105.
189. Yokoi, H., et al., *Role of connective tissue growth factor in profibrotic action of transforming growth factor-beta: a potential target for preventing renal fibrosis.* Am J Kidney Dis, 2001. **38**(4 Suppl 1): p. S134-8.
190. Grotendorst, G.R., H. Okochi, and N. Hayashi, *A novel transforming growth factor beta response element controls the expression of the connective tissue growth factor gene.* Cell Growth Differ, 1996. **7**(4): p. 469-80.
191. Holmes, A., et al., *CTGF and SMADs, maintenance of scleroderma phenotype is independent of SMAD signaling.* J Biol Chem, 2001. **276**(14): p. 10594-601.
192. Leask, A., et al., *Connective tissue growth factor gene regulation. Requirements for its induction by transforming growth factor-beta 2 in fibroblasts.* J Biol Chem, 2003. **278**(15): p. 13008-15.
193. Chen, Y., et al., *CTGF expression in mesangial cells: involvement of SMADs, MAP kinase, and PKC.* Kidney Int, 2002. **62**(4): p. 1149-59.
194. Nony, P.A., G. Nowak, and R.G. Schnellmann, *Collagen IV promotes repair of renal cell physiological functions after toxicant injury.* Am J Physiol Renal Physiol, 2001. **281**(3): p. F443-53.
195. Esposito, C., et al., *Hepatocyte growth factor (HGF) modulates matrix turnover in human glomeruli.* Kidney Int, 2005. **67**(6): p. 2143-50.
196. Liu, D.G. and T.M. Wang, *Role of connective tissue growth factor in experimental radiation nephropathy in rats.* Chin Med J (Engl), 2008. **121**(19): p. 1925-31.
197. Brigstock, D.R., *The CCN family: a new stimulus package.* J Endocrinol, 2003. **178**(2): p. 169-75.
198. Brigstock, D.R., et al., *Proposal for a unified CCN nomenclature.* Mol Pathol, 2003. **56**(2): p. 127-8.
199. Lee, S.H., et al., *Nephroblastoma overexpressed gene (NOV) expression in rat hepatic stellate cells.* Biochem Pharmacol, 2004. **68**(7): p. 1391-400.
200. van Roeyen, C.R., et al., *CCN3 is a novel endogenous PDGF-regulated inhibitor of glomerular cell proliferation.* Kidney Int, 2008. **73**(1): p. 86-94.
201. Trojanowska, M., *Role of PDGF in fibrotic diseases and systemic sclerosis.* Rheumatology (Oxford), 2008. **47 Suppl 5**: p. v2-4.
202. Floege, J., F. Eitner, and C.E. Alpers, *A new look at platelet-derived growth factor in renal disease.* J Am Soc Nephrol, 2008. **19**(1): p. 12-23.
203. Taneda, S., et al., *Obstructive uropathy in mice and humans: potential role for PDGF-D in the progression of tubulointerstitial injury.* J Am Soc Nephrol, 2003. **14**(10): p. 2544-55.
204. Seifert, R.A., C.E. Alpers, and D.F. Bowen-Pope, *Expression of platelet-derived growth factor and its receptors in the developing and adult mouse kidney.* Kidney Int, 1998. **54**(3): p. 731-46.
205. Alvarez, R.H., H.M. Kantarjian, and J.E. Cortes, *Biology of platelet-derived growth factor and its involvement in disease.* Mayo Clin Proc, 2006. **81**(9): p. 1241-57.

206. Heldin, C.H., *Development and possible clinical use of antagonists for PDGF and TGF-beta.* Ups J Med Sci, 2004. **109**(3): p. 165-78.
207. Alpers, C.E., et al., *PDGF-receptor localizes to mesangial, parietal epithelial, and interstitial cells in human and primate kidneys.* Kidney Int, 1993. **43**(2): p. 286-94.
208. Tang, W.W., et al., *Platelet-derived growth factor-BB induces renal tubulointerstitial myofibroblast formation and tubulointerstitial fibrosis.* Am J Pathol, 1996. **148**(4): p. 1169-80.
209. Bessho, K., et al., *Counteractive effects of HGF on PDGF-induced mesangial cell proliferation in a rat model of glomerulonephritis.* Am J Physiol Renal Physiol, 2003. **284**(6): p. F1171-80.
210. Uchida, K., et al., *Involvement of MAP kinase cascades in Smad7 transcriptional regulation.* Biochem Biophys Res Commun, 2001. **289**(2): p. 376-81.
211. Giannopoulou, M., et al., *Hepatocyte growth factor exerts its anti-inflammatory action by disrupting nuclear factor-kappaB signaling.* Am J Pathol, 2008. **173**(1): p. 30-41.
212. Schlondorff, D.O., *Overview of factors contributing to the pathophysiology of progressive renal disease.* Kidney Int, 2008. **74**(7): p. 860-6.
213. Vielhauer, V., et al., *Obstructive nephropathy in the mouse: progressive fibrosis correlates with tubulointerstitial chemokine expression and accumulation of CC chemokine receptor 2- and 5-positive leukocytes.* J Am Soc Nephrol, 2001. **12**(6): p. 1173-87.
214. Satriano, J.A., et al., *Regulation of RANTES and ICAM-1 expression in murine mesangial cells.* J Am Soc Nephrol, 1997. **8**(4): p. 596-603.
215. Yazawa, K., et al., *Direct transfer of hepatocyte growth factor gene into kidney suppresses cyclosporin A nephrotoxicity in rats.* Nephrol Dial Transplant, 2004. **19**(4): p. 812-6.
216. Flotte, T.R., S.A. Afione, and P.L. Zeitlin, *Adeno-associated virus vector gene expression occurs in nondividing cells in the absence of vector DNA integration.* Am J Respir Cell Mol Biol, 1994. **11**(5): p. 517-21.
217. Podsakoff, G., K.K. Wong, Jr., and S. Chatterjee, *Efficient gene transfer into nondividing cells by adeno-associated virus-based vectors.* J Virol, 1994. **68**(9): p. 5656-66.
218. McCarty, D.M., P.E. Monahan, and R.J. Samulski, *Self-complementary recombinant adeno-associated virus (scAAV) vectors promote efficient transduction independently of DNA synthesis.* Gene Ther, 2001. **8**(16): p. 1248-54.
219. Chen, S., et al., *Gene delivery in renal tubular epithelial cells using recombinant adeno-associated viral vectors.* J Am Soc Nephrol, 2003. **14**(4): p. 947-58.
220. Ito, K., et al., *Adeno-associated viral vector transduction of green fluorescent protein in kidney: effect of unilateral ureteric obstruction.* BJU Int, 2008. **101**(3): p. 376-81.
221. Mahato, R.I., *Biomaterials for Delivery and Targeting of Proteins and Nucleic Acids.* Vol. 1. 2004: CRC.
222. Kliem, V., et al., *Mechanisms involved in the pathogenesis of tubulointerstitial fibrosis in 5/6-nephrectomized rats.* Kidney Int, 1996. **49**(3): p. 666-78.
223. Inoue, T., et al., *Hepatocyte growth factor counteracts transforming growth factor-beta1, through attenuation of connective tissue growth factor induction, and prevents renal fibrogenesis in 5/6 nephrectomized mice.* FASEB J, 2003. **17**(2): p. 268-70.
224. Libetta, C., et al., *Stimulation of hepatocyte growth factor in human acute renal failure.* Nephron, 1998. **80**(1): p. 41-5.
225. Nakamura, T., et al., *Molecular cloning and expression of human hepatocyte growth factor.* Nature, 1989. **342**(6248): p. 440-3.
226. Miyazawa, K., T. Shimomura, and N. Kitamura, *Activation of hepatocyte growth factor in the injured tissues is mediated by hepatocyte growth factor activator.* J Biol Chem, 1996. **271**(7): p. 3615-8.
227. Tajima, H., et al., *Tissue distribution of hepatocyte growth factor receptor and its exclusive down-regulation in a regenerating organ after injury.* J Biochem, 1992. **111**(3): p. 401-6.

REFERENCES

228. Yanagita, K., et al., *Hepatocyte growth factor may act as a pulmotrophic factor on lung regeneration after acute lung injury.* J Biol Chem, 1993. **268**(28): p. 21212-7.
229. Nicklin, S.A., et al., *Efficient and selective AAV2-mediated gene transfer directed to human vascular endothelial cells.* Mol Ther, 2001. **4**(3): p. 174-81.
230. Grifman, M., et al., *Incorporation of tumor-targeting peptides into recombinant adeno-associated virus capsids.* Mol Ther, 2001. **3**(6): p. 964-75.
231. Engelstadter, M., et al., *Targeting human T cells by retroviral vectors displaying antibody domains selected from a phage display library.* Hum Gene Ther, 2000. **11**(2): p. 293-303.

6. Supplements

Table S2: GO-Clusteranalyses of genes down-regulated by hHGF in NRK49F cells. Genes were subdivided into the three functional clusters: molecular function, cellular component and biological process. Repetitions in genes with different fold changes represent splice variants. Only genes with a fold change greater than 2.0 were considered.

Gene Title	Gene Symbol	Fold Change
A) molecular function		
1) extracellular matrix		
tenascin N (predicted)	Tnn_predicted	25.06
chitinase 3-like 1	Chi3l1	24.58
matrix metallopeptidase 9	Mmp9	5.55
matrix metallopeptidase 9	Mmp9	5.45
thrombospondin 2	Thbs2	5.01
connective tissue growth factor	Ctgf	3.60
tumor necrosis factor receptor superfamily, member 11b (osteoprotegerin)	Tnfrsf11b	3.31
procollagen, type XII, alpha 1	Col12a1	2.94
bone morphogenetic protein 4	Bmp4	2.86
matrix Gla protein	Mgp	2.86
lumican	Lum	2.84
osteomodulin	Omd	2.79
a disintegrin-like and metallopeptidse (reprolysin type) with thrombospondin type 1 motif, 1	Adamts1	2.76
latent transforming growth factor beta binding protein 2	Ltbp2	2.73
procollagen, type 1, alpha 1	Col1a1	2.68
matrix metallopeptidase 2	Mmp2	2.63
A disintegrin-like and metallopeptidase (reprolysin type) with thrombospondin type 1 motif, 5 (aggrecanase-2)	Adamts5	2.63
tissue inhibitor of metalloproteinase 2	Timp2	2.61
ADAMTS-like 4	Adamtsl4	2.48
A disintegrin-like and metalloprotease (reprolysin type) with thrombospondin type 1 motif, 9 (predicted)	Adamts9_predicted	2.48
fibulin 5	Fbln5	2.47
secreted acidic cysteine rich glycoprotein	Sparc	2.47
periostin, osteoblast specific factor (predicted)	Postn_predicted	2.44
cysteine rich protein 61	Cyr61	2.42
secreted acidic cysteine rich glycoprotein	Sparc	2.29
procollagen, type V, alpha 2	Col5a2	2.27
ADAMTS-like 5 (predicted)	Adamtsl5_predicted	2.24
matrix metallopeptidase 13	Mmp13	2.22
procollagen, type XII, alpha 1	Col12a1	2.14
matrix metallopeptidase 12	Mmp12	2.10
matrix metallopeptidase 11	Mmp11	2.06
Matrix metallopeptidase 14 (membrane-inserted)	Mmp14	2.05
Matrix metallopeptidase 3	Mmp 3	1.58

B) cellular component
1) extracellular space

procollagen, type IV, alpha 2 (predicted)	Col4a2_predicted	2.12
procollagen, type IV, alpha 1	Col4a1	1.91

2) extracellular region part

sarcoglycan, gamma (dystrophin-associated glycoprotein)	Sgcg	5.85
similar to Beta-sarcoglycan (Beta-SG) (43 kDa dystrophin-associated glycoprotein) (43DAG)	LOC680229 /// LOC687025	1.72

C) biological process
1) cell proliferation

CD74 antigen (invariant polypeptide of major histocompatibility complex, class II antigen-associated)	Cd74	9.58
chemokine (C-X-C motif) ligand 12	Cxcl12	7.33
c-fos induced growth factor	Figf	6.90
chemokine (C-X-C motif) ligand 12	Cxcl12	6.75
chemokine (C-X-C motif) ligand 12	Cxcl12	6.28
colony stimulating factor 2 (granulocyte-macrophage)	Csf2	5.46
growth arrest specific 6	Gas6	4.44
bone morphogenetic protein 4	Bmp4	2.86
transforming growth factor, beta 2	Tgfb2	2.78
Transforming growth factor, beta 2	Tgfb2	2.69
Transforming growth factor, beta 2	Tgfb2	2.60
cysteine rich protein 61	Cyr61	2.42
growth arrest specific 6	Gas6	2.40
Matrix metallopeptidase 14 (membrane-inserted)	Mmp14	2.05
platelet derived growth factor receptor, alpha polypeptide	Pdgfra	2.04
c-fos induced growth factor	Figf	2.01

2) immune response

chemokine (C-C motif) ligand 5	Ccl5	22.76
complement component 3	C3	19.50
similar to Small inducible cytokine B13 precursor (CXCL13) (B lymphocyte chemoattractant) (CXC chemokine BLC)	LOC498335	17.90
RT1 class II, locus Da	RT1-Da	17.58
chemokine (C-X3-C motif) ligand 1	Cx3cl1	10.03
RT1 class II, locus Ba	RT1-Ba	7.75
complement component factor H	Cfh	7.02
RT1 class II, locus Ba	RT1-Ba	6.63
RT1 class II, locus Bb	RT1-Bb	5.86
colony stimulating factor 3 (granulocyte)	Csf3	5.86
peptidoglycan recognition protein 1	Pglyrp1	5.53
interleukin 6	Il6	5.31
RT1 class II, locus Db1	RT1-Db1	4.92
chemokine (C-C motif) ligand 20	Ccl20	4.88
chemokine (C-X-C motif) ligand 11	Cxcl11	4.03
myxovirus (influenza virus) resistance 1	Mx1	3.96

major histocompatibility complex, class II, DM beta	Hla-dmb	3.92
RT1 class Ib, locus S3	RT1-S3	3.77
RT1 class Ib, locus Aw2 /// RT1 class Ia, locus A2 /// RT1 class I, A3	RT1-A2 /// RT1-A3 /// RT1-Aw2	3.76
RT1 class II, locus Db1	RT1-Db1	3.74
ubiquitin D	Ubd	3.71
mannan-binding lectin serine peptidase 1	Masp1	3.58
mannan-binding lectin serine peptidase 1	Masp1	3.57
Fc receptor, IgG, alpha chain transporter	Fcgrt	3.28
RT1 class Ib, locus S3	RT1-S3	3.24
complement component 1, s subcomponent /// similar to complement component 1, s subcomponent (predicted)	C1s /// RGD1561715_predicted	3.20
cathepsin C	Ctsc	3.16
RT1 class Ib, locus S3	RT1-S3	3.06
cathepsin C	Ctsc	3.05
chemokine (C-C motif) ligand 7	Ccl7	2.98
RT1 class I, CE5	RT1-CE5	2.97
Beta-2 microglobulin	B2m	2.91
CD69 antigen	Cd69	2.68
chemokine (C-X-C motif) ligand 2	Cxcl2	2.65
guanylate nucleotide binding protein 2	Gbp2	2.58
neuraminidase 1	Neu1	2.55
Tumor necrosis factor receptor superfamily, member 6	Tnfrsf6	2.43
gene model 1960, (NCBI)	Gm1960	2.34
hemochromatosis	Hfe	2.32
Tumor necrosis factor receptor superfamily, member 6	Tnfrsf6	2.31
tumor necrosis factor (ligand) superfamily, member 13	Tnfsf13	2.24
gene model 1960, (NCBI)	Gm1960	2.15
gene model 1960, (NCBI)	Gm1960	2.11
RT1 class Ib, locus M3	RT1-M3	2.10
major histocompatibility complex, class II, DM alpha	Hla-dma	2.02
RT1 class Ib, locus S3 /// RT1-149 protein	RT1-149 /// RT1-S3	2.01

3) signal transduction

ectonucleotide pyrophosphatase/phosphodiesterase 2	Enpp2	32.52
tenascin N (predicted)	Tnn_predicted	25.06
Rho family GTPase 3	Rnd3	5.07
similar to interferon-inducible GTPase	RGD1309362	5.02
similar to Ras-related protein Rab-1B	LOC682488 /// MGC105830	4.75
platelet derived growth factor receptor, beta polypeptide	Pdgfrb	4.67
Gardner-Rasheed feline sarcoma viral (Fgr) oncogene homolog	Fgr	4.23
similar to Ras-related protein Rab-1B	LOC682488 /// MGC105830	4.09
Rho family GTPase 3	Rnd3	3.78
connective tissue growth factor	Ctgf	3.60
ephrin B1	Efnb1	3.48
bone morphogenetic protein 6	Bmp6	3.35
frizzled homolog 1 (Drosophila)	Fzd1	3.35

bone morphogenetic protein 6	Bmp6	3.33
tumor necrosis factor receptor superfamily, member 11b (osteoprotegerin)	Tnfrsf11b	3.31
receptor (calcitonin) activity modifying protein 2	Ramp2	3.30
G protein-coupled receptor 68 (predicted)	Gpr68_predicted	3.25
ras homolog gene family, member J	Rhoj	3.13
similar to Opsin-3 (Encephalopsin) (Panopsin)	LOC498289	2.82
a disintegrin-like and metallopeptidse (reprolysin type) with thrombospondin type 1 motif, 1	Adamts1	2.76
RAR-related orphan receptor alpha (predicted)	Rora_predicted	2.76
latent transforming growth factor beta binding protein 2	Ltbp2	2.73
RAR-related orphan receptor alpha (predicted)	Rora_predicted	2.71
similar to integrin, beta-like 1	LOC498564	2.67
ras homolog gene family, member J	Rhoj	2.66
signal-transducing adaptor protein-2	Stap2	2.62
tissue inhibitor of metalloproteinase 2	Timp2	2.61
guanine nucleotide binding protein (G protein), gamma 8 subunit	Gng8	2.51
Down syndrome critical region gene 1-like 1	Dscr1l1	2.49
secreted acidic cysteine rich glycoprotein	Sparc	2.47
serine/threonine/tyrosine interacting-like 1	Styxl1	2.46
Phosphodiesterase 4B, cAMP specific	Pde4b	2.42
synaptojanin 2 binding protein	Synj2bp	2.39
chaperone, ABC1 activity of bc1 complex like (S. pombe)	Cabc1	2.38
secreted frizzled-related protein 4	Sfrp4	2.37
amyloid beta (A4) precursor protein	App	2.35
RAS-like family 11 member B	Rasl11b	2.34
similar to integrin beta-5 (predicted)	RGD1563276_predicted	2.34
G protein-coupled receptor 176	Gpr176	2.32
Janus kinase 2	Jak2	2.31
heme oxygenase (decycling) 1	Hmox1	2.31
protein tyrosine phosphatase, non-receptor type 1	Ptpn1	2.30
Ras and Rab interactor 2 (predicted)	Rin2_predicted	2.29
secreted acidic cysteine rich glycoprotein	Sparc	2.29
Signal transducer and activator of transcription 2	Stat2	2.26
BMP and activin membrane-bound inhibitor, homolog (Xenopus laevis)	Bambi	2.20
carboxypeptidase E	Cpe	2.19
interferon gamma receptor 2 (predicted)	Ifngr2_predicted	2.18
Discoidin domain receptor family, member 2	Ddr2	2.18
cytokine inducible SH2-containing protein	Cish	2.18
programmed cell death 6 interacting protein	Pdcd6ip	2.16
regulator of G-protein signalling 3	Rgs3	2.12
Rho-guanine nucleotide exchange factor (predicted)	Rgnef_predicted	2.12
amyloid beta (A4) precursor-like protein 2	Aplp2	2.10
amyloid beta (A4) precursor protein	App	2.10
chemokine orphan receptor 1	Cmkor1	2.06
plasminogen activator, tissue	Plat	2.05

very low density lipoprotein receptor	Vldlr	2.05
phosphatidylinositol 3-kinase, regulatory subunit, polypeptide 1	Pik3r1	2.02

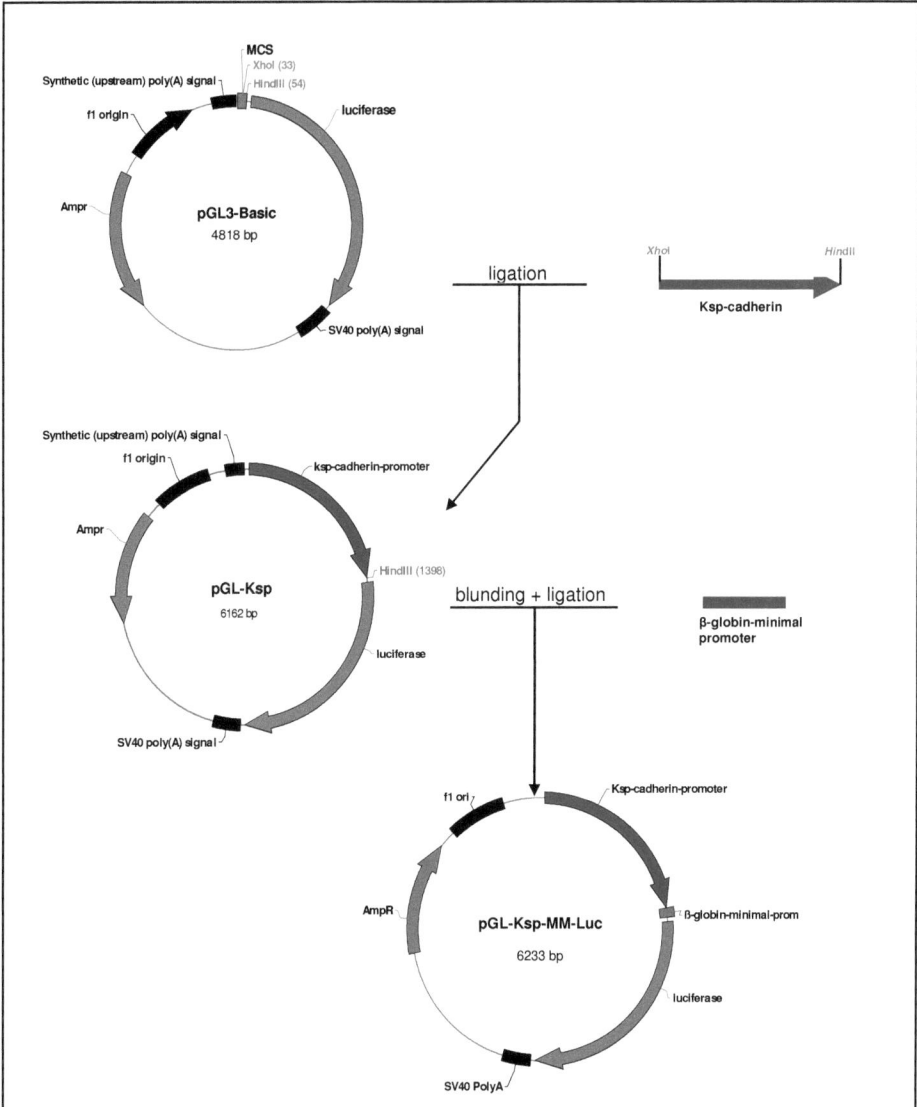

Fig. S1: Ksp-cadherin promoter enhanced by the β-globin-minimalpromotor
The reporter plasmid pGL-Ksp-MM-Luc containing upstream of the firefly-luciferase the Ksp-cadherin promotor followed by a β-globin-minimalpromotor as enhancer was generated by digestion-steps as well as PCR amplification and oligonucleotide-dimerisation. The pGL3-Basic vector (Promega, Germany) was digested with XhoI and HindIII and used as backbone. The 1342-bp region of the Ksp-cadherin promoter was amplified using specific primers (Ksp-Cad-F and Ksp-Cad-R) designed to simultaneously generate a 5' XhoI- and a 3' HindIII-restriction site. The amplification product was ligated into the pGL3-backbone resulting in the vector pGL3-Ksp. This vector in turn was opened with HindIII and the β-globin-minimalpromotor, generated by dimerisation was inserted. The resulting reporter plasmid was named pGL-Ksp-MM-Luc. The sequences of the plasmid-inserts were verified by DNA-sequencing.

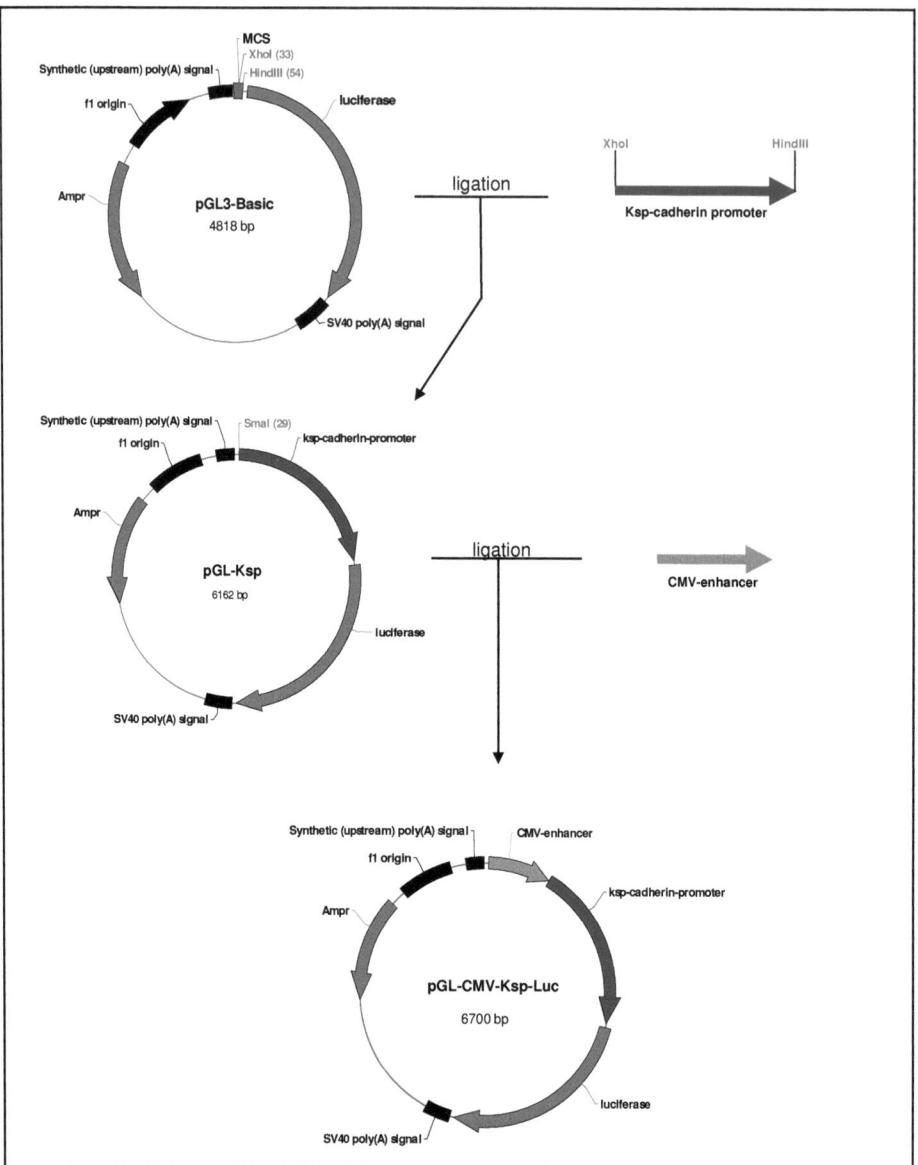

Fig. S2: Ksp-cadherin promoter enforced by a CMV-enhancer
The construction of the reporter plasmid pGL-CMV-Ksp-Luc, containing the CMV-enhancer instead of the β-globin-minimalpromoter, was similar to the pGL3-Ksp-β-globin generation. The intermediate step resulting in the vector pGL3-Ksp was the same. By dint of PCR amplification the CMV-enhancer was amplified (primer CMV-enhancer-F and CMV-enhancer-R, see 2.1.5 table 4) out of the vector pEGFP-C1 (Clontech, Germany). The PCR product was cloned into the unique SmaI site upstream of the Ksp-cadherin promoter in the pGL3-Ksp, resulting in the vector pGL-CMV-Ksp-Luc. The correct sequences of the inserts were verified by DNA-sequencing.

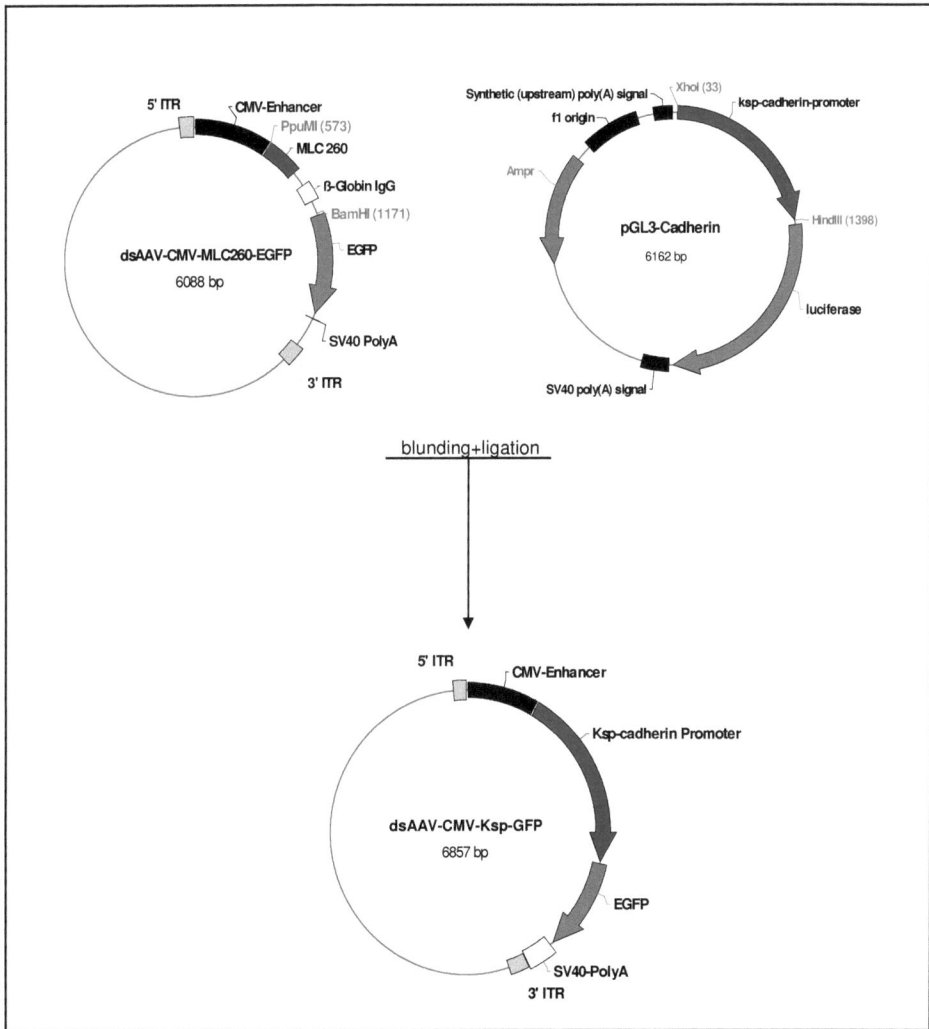

Fig. S3: Construction of the CMV-Ksp-GFP reporter construct. The reporter plasmid for the mammalian promoter Ksp-Cadherin enforced by the CMV-enhancer was constructed by replacing the region of the MLC260-β-globin IgG promoter region in the plasmid dsAAV-CMV-MLC260-EGFP by the Ksp-cadherin promoter using the flanking restriction enzymes *PpU*MI and *Bam*HI. Therefore, the resulting backbone was end-filled with T4 polymerase and the also blunded Ksp-promoter, excised out of the pGL3-Ksp via *Xho*I and *Hin*dIII, was inserted. The resulting plasmid was named dsAAV-CMV-Ksp-GFP.

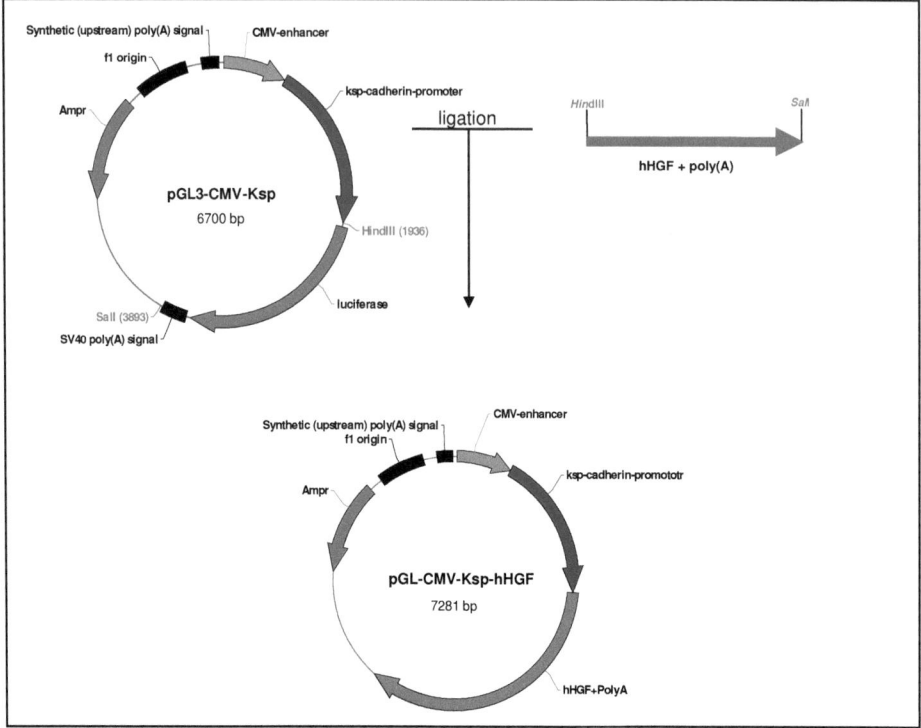

Fig. S4: Construction of a kidney specific hHGF-expression cassette
The hHGF with poly (A) was amplified out of the vector pBlast49F-hHGF (Invivogen, Germany) using primers tailed with restriction sites for HindIII and SalI (hHGF-F and hHGF-R, see 2.1.5 table 4). The pGL-Ksp (S...) was digested with HindIII and SalI to remove the luciferase and the HindIII and SalI restricted amplicon of hHGF+poly(A) was inserted, resulting in the plasmid pGL-CMV-Ksp-hHGF.

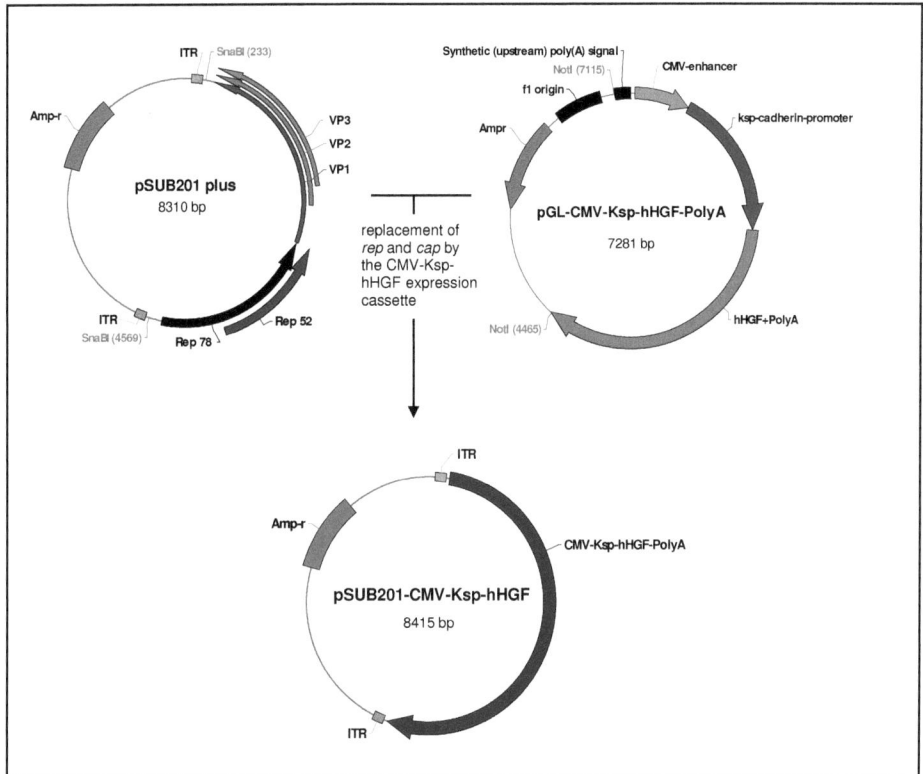

Fig. S5: Subcloning of the hHGF-expression cassette for AAV8 and 9 packaging
Restriction digest with SnaBI causes the removal of the AAV coding region leaving the AAV terminal repeats, the only *cis* acting sequences required for recombinant virus production, intact in the plasmid backbone. For the subcloning of the hHGF-cassette (3.3.1) both ORF's of the pSUB201, *rep* and *cap*, were removed by digestion with SnaBI. The hHGF expression cassette (CMV-Ksp-hHGF) was excised out of the plasmid generated above using NotI, treated with T4-polymerase and cloned into the pSUB201 backbone. The so generated transgene-containing AAV-vector was packaged into AAV8 and 9 capsids, respectively.

Die VDM Verlagsservicegesellschaft sucht für wissenschaftliche Verlage abgeschlossene und herausragende

Dissertationen, Habilitationen, Diplomarbeiten, Master Theses, Magisterarbeiten usw.

für die kostenlose Publikation als Fachbuch.

Sie verfügen über eine Arbeit, die hohen inhaltlichen und formalen Ansprüchen genügt, und haben Interesse an einer honorarvergüteten Publikation?

Dann senden Sie bitte erste Informationen über sich und Ihre Arbeit per Email an *info@vdm-vsg.de*.

Sie erhalten kurzfristig unser Feedback!

VDM Verlagsservicegesellschaft mbH
Dudweiler Landstr. 99　　　　　　Telefon　+49 681 3720 174
D - 66123 Saarbrücken　　　　　　Fax　　　+49 681 3720 1749

www.vdm-vsg.de

Die VDM Verlagsservicegesellschaft mbH vertritt

Printed by Books on Demand GmbH, Norderstedt / Germany